GN 62.8 .M37

Marks, Jonathan

Human biodiversity

DATE DUE

HUMAN BIODIVERSITY

FOUNDATIONS OF HUMAN BEHAVIOR
An Aldine de Gruyter Series of Texts and Monographs

SERIES EDITOR

Sarah Blaffer Hrdy
University of California, Davis

Richard D. Alexander, **The Biology of Moral Systems**

Laura L. Betzig, **Despotism and Differential Reproduction: A Darwinian View of History**

Russell L. Ciochon and John G. Heagle (Eds.), **Primate Evolution and Human Origins**

Martin Daly and Margo Wilson, **Homicide**

Irenäus Eibl-Fibesfeldt, **Human Ethology**

Richard J. Gelles and Jane B. Lancaster (Eds.), **Child Abuse and Neglect: Biosocial Dimensions**

Kathleen R. Gibson and Anne C. Petersen (Eds.), **Brain Maturation and Cognitive Development: Comparative and Cross-Cultural Perspectives**

Barry S. Hewlett (Ed.), **Father-Child Relations: Cultural and Biosocial Contexts**

Warren G. Kinzey (Ed.), **New World Primates: Ecology, Evolution and Behavior**

Frederick F. Grine (Ed.), **Evolutionary History of the "Robust" Australopithecines**

Jane B. Lancaster, Jeanne Altmann, Alice S. Rossi, and Lonnie R. Sherrod (Eds.), **Parenting Across the Life Span: Biosocial Dimensions**

Jane B. Lancaster and Beatrix A. Hamburg (Eds.), **School Age Pregnancy and Parenthood: Biosocial Dimensions**

Jonathan Marks, **Human Biodiversity: Genes, Race, and History**

Richard B. Potts, **Early Hominid Activities at Olduvai**

Eric Alden Smith, **Inujjuamiut Foraging Strategies**

Eric Alden Smith and Bruce Winterhalder (Eds.), **Evolutionary Ecology and Human Behavior**

Patricia Stuart-Macadam and Katherine Dettwyler, **Breastfeeding: A Biocultural Perspective**

Patricia Stuart-Macadam and Susan Kent (Eds.), **Diet, Demography, and Disease: Changing Perspectives on Anemia**

Wenda R. Trevathan, **Human Birth: An Evolutionary Perspective**

James W. Wood, **Dynamics of Human Reproduction: Biology, Biometry, Demography**

HUMAN BIODIVERSITY

Genes, Race, and History

JONATHAN MARKS

ALDINE DE GRUYTER

New York

About the Author

Jonathan Marks is Associate Professor of Anthropology, Yale University. He earned his M.S. in genetics, and M.A. and Ph.D. in anthropology at the University of Arizona, and has conducted postdoctoral research in genetics at the University of California at Davis. Dr. Marks's work on "molecular anthropology" has been widely published in professional journals.

Copyright © 1995 Walter de Gruyter, Inc., New York
All rights reserved. No part of this publication may be reproduced or transmitted in any form or by any means, electronic or mechanical, including photocopy, recording, or any information storage or retrieval system, without permission in writing from the publisher.

ALDINE DE GRUYTER
A division of Walter de Gruyter, Inc.
200 Saw Mill River Road
Hawthorne, New York 10532

This publication is printed on acid free paper ∞

Library of Congress Cataloging-in-Publication Data
Marks, Jonathan (Jonathan M.)
    Human biodiversity : genes, race, and history / Jonathan Marks.
        p. cm.—(Foundations of human behavior)
    Includes bibliographical references and index.
    ISBN 0-202-02032-0 (cloth : alk. paper).—ISBN 0-202-02033-9
(pbk. : alk. paper)
    1. Physical anthropology.   2. Human population genetics.
3. Biological diversity.   4. Molecular genetics.   I. Title.
II. Series.
GN62.8.M37   1994
573—dc20   94-19450
                                                                        CIP
Manufactured in the United States of America

10 9 8 7 6 5 4 3 2 1

For my parents

Contents

Acknowledgments

My sincerest gratitude goes to Sarah Blaffer Hrdy for suggesting and encouraging this project, and for her ever-insightful comments on drafts at various stages. George Hersey also read and offered his valuable advice on the manuscript from top to bottom.

For discussion of, comments on, and/or help with various parts of the manuscript, I thank Amos Deinard, Rebecca Fisher, Alan Goodman, Jim Moore, Deanna Petrochilos, Alison Richard, Vincent Sarich, Michael Seaman, Mark Stoneking, Karen Strier, and Alan Swedlund.

Special thanks also to my editor at Aldine, Richard Koffler; and to the people who saw the book through its production phase, especially Arlene Perazzini.

I wish to thank the copyright holders for permission to use the following illustrations, which I adapted for use in this book: Historical Library, Cushing-Whitney Medical Library, Yale University, for the Tulpius and Tyson "Orang-Outangs" (both chapter 1); Sterling Memorial Library, Yale University, for the portraits of Buffon and Linnaeus, and the University Museum, University of Pennsylvania, for the photograph of Carleton Coon by Reuben Goldberg (both chapter 3); the Bettmann Archive for the photograph of Franz Boas (chapter 4); the National Academy of Sciences for the photograph of Charles Davenport from *Biographical Memoirs*, Volume 25, copyright 1948 (chapter 5); the Peabody Museum of Archaeology and Ethnology, Harvard University, for the photograph of Earnest Hooton (chapter 6); Mrs. Prue Napier and the Napier estate for the adaptation of the drawing of the human and chimpanzee hands by John Napier, first published in 1980 (chapter 10); and Harvard University Press for the adaptation of one of the ant legs drawn by Sarah Landry for *Insect Societies* by Edward O. Wilson, copyright 1971, and *Scientific American* for the adaptation of the anatomical drawing of a human leg by Enid Kotschnig from "The Antiquity of Human Walking" by John Napier, April 1967, copyright 1967 (together in chapter 12).

I would also like to acknowledge the sources of other illustrations in these pages: Unione Tipografico-Editrice Torinese and Gustav Fischer Verlag for the pair of photographs from Renato Biasutti, *Le razze e i popoli della terra,* Tllild Edition, UTET, 1959 in turn retouched from R. Martin, *Lehrbuch der Anthropologie,* G. Fischer, 1914 (chapter 10); and the Antikenmuseum of the Staatliche Museen Preußischer Kulturbesitz, Berlin, for the variation on the marble bust of Cleopatra VII in its permanent collection (chapter 9). Aficionados of A.S. Romer will recognize the adaptation of the "pineapple" in chapter 7. Finally, thanks are due the copyright holder, Harcourt Brace and Company, for generously allowing me to adapt two illustrations of my own (the hominid skulls in chapter 2 and skull shapes in chapter 7) from that classic, *Evolutionary Anthropology* by Staski and Marks, published in 1992.

1

The Hierarchy

Edward Tyson's work and Charles Darwin's work were the cornerstones of a new view of the place of the human species in the natural world. A nested hierarchy of creatures exists, and we are (progressively more exclusively) primates, anthropoids, catarrhines, and hominoids. The theory of evolution explains why that hierarchy exists, because more inclusive groups share more distant ancestors. Major conceptual revolutions occurred in parallel in 19th century biology (undermining anthropocentrism) and in 20th century anthropology (undermining ethnocentrism) in understanding our place in the world.

INTRODUCTION

Are humans unique?

This simple question, at the very heart of the hybrid field of biological anthropology, poses one of the falsest of dichotomies—with a stereotypical humanist answering in the affirmative and a stereotypical scientist answering in the negative.

Any zoologist is forced to concede that humans *are* unique in certain ways—that we are not apes, and are easily distinguished from apes, in the same way that ducks are not pigeons, and lions are not wolves. It is a possibly trivial sense—simply the observation that in the panoply of nature, our species is our species and not some other species—that implies at least a minimal amount of distinctness.

We are *not* unique in our fundamental biology, however. Our cells are almost indistinguishable from an ape's cells—and as different from the cells of one ape species as that ape's cells may be from those of another ape species. The components of our bodies, their functions and processes, are exceedingly similar to an ape's. And one has only to watch a group of chimpanzees interacting to sense that their minds are like our minds.

Ape biology and human biology are of a piece with one another. And ultimately of a piece with clam biology and fly biology.

The study of human biology, however, is different from the study of the biology of other species. In the simplest terms, people's lives and welfare may depend upon it, in a sense that they may not depend on the study of other scientific subjects. Where science is used to validate ideas—four out of five scientists preferring a brand of cigarettes or toothpaste—there is a tendency to accept the judgment as authoritative without asking the kinds of questions we might ask of other citizens' pronouncements. Why, after all, would it matter what four out of five scientists prefer, unless there was some authority that came with that preference?

We can call this scientism: the acceptance of the authority of scientists. It is different from science, the process by which we come to understand and explain natural phenomena. The reason for this difference is simple. Science is the way in which we examine and confront the many things that might be true and prune them down to the few things that probably are true. It occurs by a process of "conjecture and refutation" (Karl Popper)[1] or more euphoniously, "proposal and disposal" (Peter Medawar).[2]

The paradoxical flip-side to science is that the vast majority of ideas that most scientists have ever had have been wrong. They have been refuted; they have been disposed of. Further, at any point in time, most ideas proposed by most scientists *will* ultimately be refuted and disposed of. While this is fundamentally how our knowledge of the universe grows, it has the ultimate effect—and a threatening one—of impeaching the authority of scientists. Science, in other words, undermines scientism.

Nowhere is this paradox more evident than in the study of human biological variation. Scientists' ideas are formed partly through what we like to imagine is the objective analysis of data; but also, like the ideas of anyone else, formed partly by their cultural upbringing and life experiences. The pronouncements of scientists on human variation may be as loaded with cultural prejudices as those of anyone else—and as history shows us, indeed they usually have been. Except that, as the pronouncements of scientists, these ultimately cultural values would subsequently be vested with the authority of science. The culture can consequently produce the values that the scientist validates, thus proving that the culture was right all along.

The study of biological variation in the human species is thus a bit different from other kinds of scientific endeavors. Biologists studying fruitflies certainly have the same cultural prejudices of an era and class, yet it is generally difficult to imagine those cultural prejudices pervading their work. And it is more difficult to imagine those prejudices in their work as the basis of scientific authority to oppress or to degrade the lives of other people. The English mathematician G. H. Hardy described his attraction for his own field of scientific research: "This subject has no practical use: that is to say, it cannot be used for promoting directly the destruction of human life or for accentuating the present inequalities in the distribution of wealth."[3]

Anthropology, on the other hand, can and has been used in precisely those ways. Anthropologists, consequently, are absorbed in their intellectual history—in learning from the mistakes of earlier generations of scholars. The more we understand those conceptual errors, which usually are visible only in hindsight, the more the science of the human species can grow—by the very process of proposal and disposal by which science functions.

This thesis forms the backbone of the present book. It is about the current state of our understanding of genetic diversity, its patterns and its significance, in our species. It is about the ideas that shape contemporary thought about human genetics. But the present was formed in the past; and to know where we are, it helps to know where we've been—for in some respects we are still there. One, after all, ignores intellectual history at one's own peril. But in this case one ignores the intellectual history of human diversity not only at one's own peril, but at the peril of many people.

Thus the sciences and the humanities fuse in the study of human biological diversity. The subjects are (on the one hand) data, and (on the other) the cultural history surrounding the collection and interpretation of those data. We try neither to exalt nor to profane the human species; we handle science in the same way. The human species is both different from, and similar to, other species; and science has been both useful and tragic in approaching these questions.

PATTERN AND PROCESS

The relationship of humans to the natural world is a philosophical question of long standing. In the year 1699 it became an empirical question as well. In that year Edward Tyson, the leading anatomist in England, published the results of his dissection of a chimpanzee. Tyson had already written definitive monographs on the anatomy of a dolphin and an opossum, but the subject of the new monograph was different, for it bore directly upon the place of humans in the natural sphere.

The new monograph was called "Orang-Outang, *sive* Homo Sylvestris: Or, the Anatomy of a Pygmie Compared with that of a Monkey, an Ape, and a Man." The specimen was neither an orang-utan nor a pygmy, but an infant male chimpanzee that had died of a jaw infection in

England following a fall on board ship during the voyage from West Africa.

The creature proved to be immensely interesting, not least of all because reports of such beings from remote continents tended to confuse zoology, anthropology, and mythology. There were different kinds of animals in Asia and Africa, but there were also different kinds of people, and the reports of both were being spread by travelers with, like everyone, vivid imaginations. In fact, much of the confusion would not be sorted out for a century and a half; but Tyson managed to take the first steps in that direction.

An ape had been described superficially by a Dutch anatomist named Nicolaas Tulp in 1641, but though he said he believed it came from Angola and had black hair, he nevertheless also called it an "Indian Satyr" and discussed what the natives of Borneo thought of it.[4] Tulp's account is not only highly mythological, but also unclear as to whether his subject was a chimpanzee or an orang-utan. His illustration is ambiguous (Figure 1.1). Nevertheless, Tulp named the animal following the local (Bornean) designation: *orang-outang*, or man of the woods [*orang-outang, sive homo sylvestris*].

Tyson was more secure about the origin of his subject, had seen it alive, had studied its body upon its death, and had devoted an entire monograph to it, not simply a few paragraphs in a medical text, as Tulp had done. As Tulp had reported, the animal indeed bore an extraordinary likeness to the human species.

Tyson's study showed him that there were 48 ways in which his "Pygmie" more closely resembled a human than a monkey, but only 34 ways in which it more resembled a monkey (Figure 1.2).[5] According to Tyson's biographer, Ashley Montagu, it represented the first scientific presentation "that a creature of the ape-kind was structurally more closely related to man than was any other known animal."[6]

Monkeys had been known since antiquity: venerated by the Egyptians, dissected by the Greeks. Indeed, Vesalius in Renaissance Italy demon-

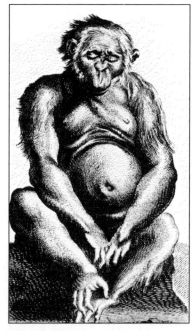

Figure 1.1. Tulp's "Orang-Outang" of 1641.

strated that in certain ways the classical Greek anatomy of Galen was based on a monkey's, rather than a human's, body.[7] Thus the similarity between human and monkey was well established, but there was certainly no doubt about the latter's being a dumb brute, an animal. The "Pygmie," on the other hand, was more ambiguous.

In spite of the extraordinary degree of similarity of organ, muscle, and bone to a person, the "Pygmie" nevertheless neither spoke nor walked. Tyson explained this paradox: the "Pygmie" didn't speak since, possessing the physical faculties, it still lacked the mental ones (which proved, as Descartes had recently argued, that mind and body are separate entities). Further, it walked in the most curious way—on all fours, but with its weight born by the knuckles of the forelimb. This was so unnatural a posture, reasoned Tyson, that it must have been walk-

Figure 1.2. Tyson's "Pygmie" of 1699.

ing that way because of its illness, for it was clearly built for good old bipedalism.

As any scientist does, Tyson used the mindset of his times to interpret his work, and that paradigm was the Great Chain of Being.[8] In other words, the "Pygmie" formed a missing link that tied humans to other creatures physically, if not intellectually. The Great Chain of Being, which figures prominently in the study of humans, subsumed three related principles:[9] first, that every species that could exist did exist; second, that every existing species could be organized along a single dimension, a line; and third, that every species on that line graded imperceptibly into the species above it and below it. Obviously different versions of this theory were adopted by individual scholars, but they all shared to some extent these postulates.[10]

The Great Chain of Being was a 17th-century interpretation of the pattern of nature, the organization one encounters upon examining the diverse forms of life. There was a parallel interpretation for how that pattern came to be, the process that generated it. The process was the instantaneous origin of each species, as is, at the beginning of history. Both the process and the pattern were miraculous, in that neither could be explained or understood by rational means. The creation and the Great Chain of Being could certainly be inferred, but how they happened or came to happen was neither known nor probably knowable.

Ultimately it was Linnaeus in the mid-18th century who overthrew the Great Chain of Being as the pattern of nature, replacing it with a "nested hierarchy"; and it was Charles Darwin in the mid-19th century who overthrew creationism with the process of "descent with modification." The replacement of the older pattern and process with the newer, and its relation to the position of humans in the scheme of things, was arguably the major conceptual innovation of the 19th century.

THE PATTERN: LINNAEUS

We attribute to Carl Linné (whose name was Latinized as Carolus Linnaeus) the initial perception of the pattern now recognized as absurdly conspicuous in nature. He is consequently hailed as the "father of systematics" by virtue of his relationship to Mother Nature.

Linnaeus is often depicted as a classifier, forever insisting on giving names to species—which of course he did, but which others had done before him. His lasting contribution, however, lay in apprehending how those named entities fit together. In other words, he found structure at a fundamental level in the natural world, but it was a different structure from the one-dimensional ranking of the Great Chain of Being. Rather, Linnaeus saw a two-dimensional pattern: a horizontal as well as a vertical dimension (Figure 1.3).[11]

He noted, for example, that of the animal species on earth, only a restricted fraction shared a fundamental similarity: the ability to nurse young. He designated these the Class Mammalia, as opposed, for example, to the Class Reptilia. Within the Class Mammalia, there were other more restricted groups that shared fundamental similarities, for example, the Order Rodentia, the Order Carnivora, and the Order Primates.[12] The last were distinguishable, for example, in having nails generally where species in other orders had claws. And within the Order Primates, there were even more restricted groups, each of which he called a genus.

The novelty was in discerning categories of equal rank at each of a series of levels. In other words, there were several classes of animals, but none was "higher" than any other, for they were all classes. Classes were a higher rank than orders, but again within any class of animals there were several orders, each of which was at the same level as every other

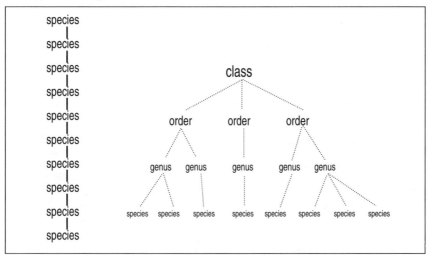

Figure 1.3. (Left) The Great Chain of Being, a one-dimensional hierarchy. (Right) Linnaeus's nested two-dimensional hierarchy.

order. Ultimately, therefore, every species was a member of a genus, a member of an order, and a member of a class. The problem was that this was incompatible with the viewpoint of the Great Chain, whereby species were not so much members of ever more inclusive clusters, but rather were simply higher or lower than any other given species.

THE OPPOSITION: BUFFON

The pattern uncovered by Linnaeus, the nested hierarchy of life, did not imply that an evolutionary process had generated it. Linnaeus, until near the end of his own life, maintained that throughout the history of life, there had been no new species formed. Only in his old age did he concede that hybridization between two species could in fact create a new species.[13]

Linnaeus's views were opposed throughout his lifetime by the Count de Buffon, a French naturalist who had a popular following, and addressed philosophical and scientific questions with equal gusto.[14] As we will see in Chapter 3, Linnaeus and Buffon differed in their approach to the human species in nature. To Buffon, there were no categories higher than the species (Figure 1.4). Species were the units that composed the Great Chain, and grouping them in any other manner was just an artifice. In addition (and somewhat paradoxically, as Buffon had formulated an early scientific version of microevolution), Buffon felt that higher categories might imply something insidious. He considered the donkey and the horse, which Linnaeus had quite naturally placed together.

> One could attribute the slight differences between these two animals to the very ancient influence of climate, nutrition, and the fortuitous succession of many generations of small, partially degenerated wild horses. Little by little they would have degenerated so much that they would ultimately have produced a new and constant species. . . . Do the donkey and horse come originally from the same source? Are they, as the taxonomists say, of the same *family*? Or are they, and have they always been, different animals? . . .
>
> Consider, as M. Daubenton has said, that the foot of a horse, superficially so different from the hand of man, is nevertheless composed of the same bones; and that we have at the tips of each of our fingers the same horseshoe-shaped bone which terminates the foot of the animal. . . .
>
> From this point of view, not only the donkey and horse, but as well man, apes, quadrupeds, and all the animals could be regarded as constituting the same *family*. But must one conclude that within such a great and numerous family, which was called into existence from nothing by God alone, there were other smaller families, projected by nature and produced by time, some of which comprise but two individuals (like the horse and donkey), others of more individuals (like the weasel, ferret, martin, polecat, etc.). . . . And if it is once admitted that there are families of plants and

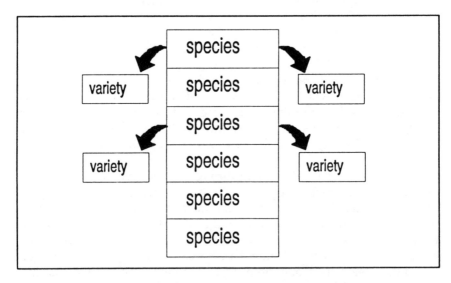

Figure 1.4. The micro-evolutionary degeneration of species, envisioned by Buffon.

animals, that the donkey is of the horse family, and that it differs only because it has degenerated, then one could equally say that the ape is a member of the human family, that it is a degenerated man, that man and ape have had a common origin like the horse and donkey—that each family among the animals and plants has had but a single stem, and that all animals have emerged from but a single animal which, through the succession of time, has produced by improvement and degeneration all the races of animals.

The naturalists who establish so casually the families of plants and animals do not seem to have grasped sufficiently the full scope of these consequences, which would reduce the immediate products of creation to a number of individuals as small as one might wish. For if it were once proved that these families could be established rationally—that of the animals and vegetables there were, I do not say several species, but only one, produced by the degeneration of another species—if it were true that the donkey were but a degenerated horse—then there would be no limits to the power of nature. One would then not be wrong to suppose that she could have drawn with time, all other organized beings from a single being.

But no: it is certain, from revelation, that all animals have participated equally in the grace of creation, that the pair of each species and of all species emerged fully formed from the hands of the Creator. . . . [15]

Buffon went on at length to debunk the theory of evolution, which did not yet exist. In other words, Buffon opposed the Linnaean classificatory system because he felt it directly implied macroevolution as a process, which could not possibly be right. Even though Linnaeus himself did not espouse such an idea, it was (according to Buffon) simply because he had not "grasped sufficiently the full scope" of the implications of his system.

Buffon maintained that species had remained stable since their formation (which he suggested might have been far earlier than his contemporaries maintained). *Within* any species, environmental conditions could well have caused populations to become distinct from one another; but certainly not to become another species.

The early years of the 19th century saw widespread acceptance of the Linnaean hierarchy as a framework for ordering Nature. Where macroevolutionary ideas surfaced (such as in the writings of Lamarck), they invariably occurred within the context of the Great Chain: New taxa were seen as emerging upward, not outward; getting better, not more diverse; climbing, rather than dividing.

And yet naturalists acknowledged that the pattern Linnaeus discerned was fundamentally right. By the early part of the 19th century, the notion of the Great Chain as an organizing principle of zoology had been dispatched, largely through the influence of Cuvier. Cuvier was certainly the leading zoologist of his day, the leading authority on ver tebrate paleontology and comparative anatomy. His studies led him forcefully to the conclusion that there was no Great Chain: whatever order existed among animals involved an organizing principle of nested categories of equal rank. How those taxa came to be, Cuvier cared not to speculate, though he never entertained the notion of species diverging and transforming through time.

Cuvier is best remembered now for his advancement of "catastrophism," an explanation for the succession of species through geological strata that invoked (obviously) catastrophes at the boundaries between geological strata, and autochthonous origins of new species by unknown (presumably divine) mechanisms.[16] Nevertheless, processes were not Cuvier's main concern—the patterns generated by the processes were his concern, as they had been for Linnaeus. And he did not see any reasonable way to place all of nature's bounty along a single line.

Cuvier knew there were at least four lines of animals: vertebrata, mollusca, articulata, and radiata. Within vertebrata there were four subdivisions as well: mammalia, reptilia, aves, and pisces. Within mammals there were nine subdivisions. Unlike Linnaeus, Cuvier split humans off from the other primates, calling humans "Bimana" and the others "Quadrumana"—two-handed or four-handed.

One could imply linearity in such a scheme by listing the "best" taxa first—and indeed, in Cuvier's *Animal Kingdom* it is his own species that is discussed first, and the lowly *Volvox* last.[17] Yet there was only a weak implication, at best, that their order of presentation (bimana, quadrumana, "carnaria" [including cheiropterans, insectivores, and carnivores], marsupials, rodents, edentates, pachyderms, ruminants, and cetaceans among the mammals) reflected an underlying linearity in their relationships to one another.

THE PROCESS: LAMARCK

Lamarck saw classification as a sterile enterprise and sought the underlying mechanisms of life. As did Buffon before him (Lamarck began his career as tutor to Buffon's son), Lamarck denied the fundamental reality of taxonomic categories, maintaining their artificiality. Indeed he went further than Buffon, and denied even that species were natural groups!

> We may, therefore, rest assured that among her productions nature has not
> really formed either classes, orders, families, genera, or constant species,

but only individuals who succeed one another and resemble those from which they sprung.[18]

Just as Cuvier retained some elements of linearity in his view of nature, Lamarck did recognize at least a little branching. Thus, reptiles gave rise to both mammals and to birds independently, he maintained—though he still saw mammals as "higher" than birds, and monotremes (like the platypus and echidna) as somehow intermediate between the two.

The suggestion that these major groups of animals were linked genealogically was Lamarck's most original contribution. By arguing against the reality of these groups of organisms, he shifted attention to the life and germination of the individual creature itself. In brief, nature was continually producing "low forms" of life, i.e., spontaneously generating. At any point in time, a creature could be challenged or stimulated by its environmental circumstances. Responding in some behavioral or anatomical manner—it would need to respond or change in order to survive—the organism would pass the modification to its offspring, and so enable its descendants to be a little more perfect. This improvement in the face of environmental challenge permitted the various primordial life-forms to rise up the scale of perfection.

To scientific readers, particularly in the empiricist tradition of English science, Lamarck's work was vain and speculative.[19] Worse, it was at the fringe of science—after all, improvement implied direction, direction implied a destination or goal, and that in turn implied some form of metaphysical map charting the course of evolution.

On the other hand, those interested principally in processes of nature were drawn inexorably to evolution, the transformation and divergence of species. First Lamarck, and by the 1850s notably Robert Chambers and Herbert Spencer, ascertained that historical biology had to involve the mutability of species.[20] They all, however, rooted their evolutionism securely in the Great Chain. To them, evolution was synonymous with progress—and progress is unidirectional. Thus, by the mid-19th century, naturalists had discerned a pattern (nested hierarchy) wedded to an indefensible process (special creation), and had discerned a process (evolution) wedded to an indefensible pattern (the Great Chain).

THE SYNTHESIS: DARWIN

Darwin's lasting contribution lay in his ability to reconcile the apparent pattern with the apparent process. He divorced the idea of evolution from the idea of progress, and therefore wrote not about the *improvement* of species, but about their *origin.* Contrary to the previous generation of evolutionists who had imagined a species to "evolve" by climbing a notch up the scale of life, Darwin modeled his version of evolution on the diversity of forms established by animal breeders. By selecting organisms with particular characteristics to be the progenitors of future generations, animal breeders had created an immense diversity of breeds with their own distinctive attributes. None was better or worse in a grand sense than any other, but all had distinctions and peculiarities. And they had accomplished this during the brief span of human history. Darwin reasoned that nature, over the vast expanse of time, could probably accomplish the same thing. Nature could select, through the environment, organisms with certain attributes, and give them a better-than-average chance of reproducing. Succeeding generations would therefore come to resemble those reproducers, since they were founding those generations, they would resemble less the average organism from the original population, and much less an organism in another environment favoring other attributes

Darwin's principle of "natural selection" not only gave a materialistic mechanism to evolution, which earlier had relied upon nebulous internal drives, or external plans, but as well afforded a means of generating diversity, rather than simply generating improvement. In other words, Darwin provided a means for the side-by-side existence of two different species that were neither better nor worse than one another—which was precisely what the Linnaean pattern required but lacked. The "natural system" of the classifiers was a pattern produced by the process of evolution, and so Darwin argued forcefully

> that the natural system is founded on descent with modification; that the characters which naturalists consider as showing true affinity between any two or more species, are those which have been inherited from a common parent, and, in so far, all true classification is genealogical; that community of descent is the hidden bond which naturalists have been unconsciously seeking, and not some unknown plan of creation.[21]

THE PLACE OF HUMANS IN NATURE

The creature Edward Tyson examined is now considered to be a member of the genus *Pan*, a chimpanzee. Along with the genus *Gorilla*, it represents the closest living relative of our own genus, *Homo*. Somewhat more distantly removed from the human-chimp-gorilla triad is the orang-utan (*Pongo*) of southeast Asia, and more distant still is the Asian gibbon (*Hylobates*) (Figure 1.5).

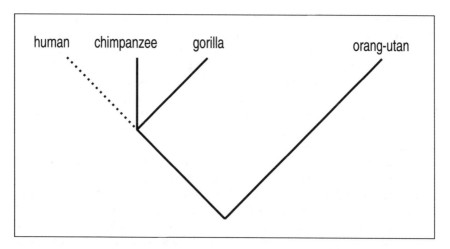

human   chimpanzee   gorilla   orang-utan

Figure 1.5. Relations of the great apes and humans. Though humans, chimps, and gorillas shared a recent ancestor, the human lineage (indicated by a dashed line) has undergone extensive change in a short period of time.

How do we know this? Two major kinds of data can be adduced in support of the branching relationships of the primates. Classical systematics derives its inferences from the kinds of information that Linnaeus and Darwin used: the anatomy and physiology of organisms. Molecular systematics takes advantage of the great technological strides in biochemistry and molecular biology to compare the hereditary material in different species.

Cross-cutting these two kinds of data are two kinds of analyses: *phenetic* and *cladistic.* In a phenetic study, we ask how species are similar to or different from one another. To a first approximation, obviously, more similar species will be more closely related. However, a large part of evolution consists of divergence—and a very divergent taxon is not likely to be very similar to anything. Thus, a species that has changed a lot in a little bit of time will look different from everything, though it may be especially closely related to another particular species. An example of this is our own genus, *Homo,* which has changed anatomically and behaviorally very much from its close relatives, the apes. Genetically, however, it was shown in the early 1960s that humans fall in neatly with the chimpanzees and gorillas.[22] The genetic changes do not accumulate at the same pace as the anatomical changes, and thus the extreme *anatomical* divergence of humans could not mask their close genetic relationships to the African apes.

Cladistic (or phylogenetic) analysis, on the other hand, focuses on the evolutionary process not so much as the accumulation of "distance" between species, which can be abstractly quantified, but rather as the historical succession of particular traits through time. A characteristic such as canine tooth size is small in humans but large in all other catarrhine primates (monkeys and apes of the Old World). We therefore hypothesize that humans evolved from ancestors that had, like other catarrhines, large canines. Further, if we encounter other catarrhines with small canines, such as those from East Africa about 1.5 million years ago, we can hypothesize that they are more closely related to us than to other primates (Figure 1.6).

The reasoning is fairly straightforward. Evolution involves changes of something into something else. The "something" is known as a plesiomorphic or primitive state; the "something else" is an apomorphic or derived state. By studying the distribution of evolutionary events, apomorphies, we analyze the patterns of the history of life. Since it is easier not to evolve than to evolve, apomorphies are relatively rare. Thus, if we encounter the same apomorphy in more than one species, it is likely that the species are closely related, for they presumably inherited that apomorphy from a recent common ancestor.

Evolutionary changes are distributed in two ways: across several species, and in a single species. Those changes in a single species are known as autapomorphies, and reflect the divergence of that species from its close relatives, for they are unique. These define its evolutionary individuality, the manner by which this species is different from oth-

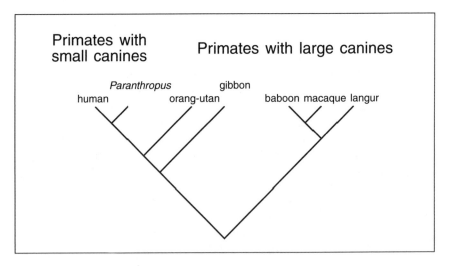

Figure 1.6. Humans and the fossil genus *Paranthropus* are linked by sharing reduced canines, a synapomorphy. The distribution of large canines fails to reveal the complex relations among the species that retain it.

ers. Changes distributed across species are known as synapomorphies, and reflect evolutionary modifications in a species that were subsequently inherited by its descendant taxa. These define clades, clusters of closest relatives. The nested hierarchy of Linnaeus, therefore, represents to a large extent the nested distribution of synapomorphies across increasingly restricted groups of species.

The key problem in cladistic analysis is the determination of polarity: Did large canines evolve into small canines, or was it vice versa? We hypothesized that large canines is the ancestral state and small canines represents the derived state, and it is concordant with other molecular and anatomical data. Thus, the category "all catarrhines with small canines" denotes a fairly exclusive group, one that consists of species closely related to one another, descended from a common ancestor in which the canines had diminished in size. The other category, "all catarrhines with large canines," doesn't tell us which groups of species are closely related: that trait is *symplesiomorphic* (a shared primitive feature), and its distribution simply gives us a number of different lineages in which the evolutionary event of interest did *not* occur.

One major problem in phylogenetic analysis involves deciding which is the derived and which is the primitive state of the feature in question. How sure can we be that the evolutionary event was really a reduction of the canines in the common ancestor of some species, rather than growth of canines in the common ancestor of others? The simplest answer is given by out-group analysis (Figure 1.7). One examines a

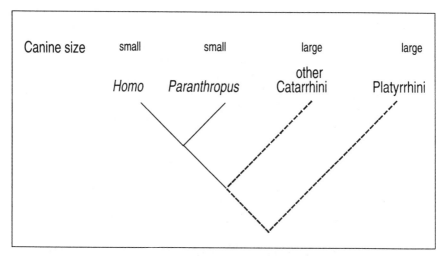

Figure 1.7. Using an out-group comparison helps to establish the polarity of an evolutionary change.

species just outside the evolutionary event in question, so that it can be
safely assumed that the evolutionary event did not affect that species.
Therefore the out-group presumably expresses the primitive trait. In this
way, the evolutionary polarity of the traits can be established, at least to
a first approximation.

In this example, the relevant out-group is the New World monkeys, or
Platyrrhini, which are rather distantly related to the group subsuming
apes and humans, within which a change in the canine teeth occurred.
The Platyrrhini have large canine teeth. This implies that the ancestral
state was large canines, and therefore that the evolutionary event under
consideration was indeed a change from large to small canines. This
consequently links *Homo* and *Paranthropus* as close relatives.

The other major problem in phylogenetic analysis is *homoplasy*, or parallel evolution, the acquisition of similar traits in different evolutionary
lines. Often species adapting to similar environments will adapt in similar ways. Or sometimes the changes are not necessarily adaptations at
all, but simply part of the possible range of forms that an organism can
take, and two species both happen to have it. Ideally, homoplasy can be
discerned by yielding a pattern that is at odds with the distribution of
other synapomorphies (shared, derived traits). For example, two genera
of primates have lost their thumbs: *Ateles* (the spider monkey) and
Colobus (the colobus monkey). Are they close relatives whose recent
common ancestor lost its thumb? The spider monkey is a platyrrhine
New World monkey; the colobus is a catarrhine Old World monkey, and
the distribution of their many other features shows them to be fairly distantly related. The loss of the thumb is consequently interpreted as two
autapomorphies in different lines, not one synapomorphy in a unique
ancestor of the two monkeys.

We localize humans among the mammals by virtue of a large suite of
synapomorphies, including: a four-chambered heart, warm-bloodedness,
hair, sweat glands, live-birth, baby teeth, a jaw composed of a single
bone, and, of course, lactation. We localize humans among the primates
by virtue of synapomorphic features of their hands and skulls, including grasping fingers, fingernails, opposable thumbs, and extensive bony
protection for the eyes—all synapomorphies that we share with other
modern primates.

Among the primates, humans share an even more exclusive group of
characteristics with gibbons, orangutans, chimps, and gorillas. These
synapomorphies include an appendix, absence of a tail, a flexible shoulder, and numerous other specializations of the trunk and upper limbs
that appear to be associated with locomotion by hanging, climbing, or
swinging.

In each case, the synapomorphies that distinguish the clade (those sets of closely related taxa) stand in contrast to the symplesiomorphies possessed by the out-groups, more distant relatives of the taxa under consideration (Figure 1.8).

The patterns of genetic relationship closely match those of anatomical relationship. For example, over a stretch of DNA spanning some of the major genes that code for hemoglobin, a strict genetic comparison can be made between various catarrhine species. Here human, chimpanzee, and gorilla all differ from one another by less than 2 percent. Each of these differs from the DNA of the orang-utan by about 3.5 percent. These four differ from the corresponding DNA of a gibbon by about 4.5 percent, and all the hominoids differ from the cercopithecoids by 7 to 8 percent. A technique called DNA hybridization, which compares a much larger portion of the genetic material more crudely, gives nearly identical results.[23]

Cladistic or phylogenetic analyses of molecular genetic data give a concordant picture as well. The relatively few detectable genetic differences among the hominoids generally map onto the relationships already given. For example, in the beta-chain of hemoglobin, the genetic instructions present in cercopithecines, gibbons, and orang-utans call for the amino acid glutamine; but the genetic instructions of humans, chimps, and gorillas call for the amino acid proline. Again, the interpretation is that the substitution of proline at position 125 of beta-hemoglobin is a synapomorphy, and the retention of glutamine is a symplesiomorphy in the other taxa.[24]

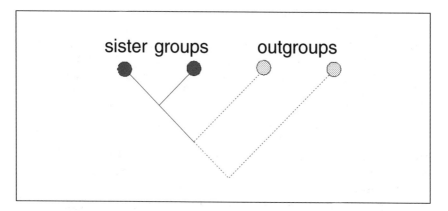

Figure 1.8. Closest relatives (sister groups) share synapomorphies; out-groups share symplesiomorphies.

ANCHORING THE EMERGENCE OF HUMANS

Thus modern theory and technology uphold the fundamental insights of Linnaeus and Darwin. Humans lie biologically among the mammals, primates, and apes, reflecting a common ancestry shared increasingly more recently and more exclusively with other mammals, other primates, and other apes.

And yet humans are unique. Like any species, humans have a suite of uniquely derived features—autapomorphies—which distinguish us from the "other" apes. Although humans are apes in fundamental genetic, anatomical, and social senses, they are also different from the apes. The nature of those differences, as well as those similarities, defines what it means to be human. To suggest that humans are "nothing but" apes—as is sometimes done in a rather perverse scientific manner—is to miss the basic importance of autapomorphies in defining what a species is. Humans have many autapomorphies, and their elaboration and significance is where the sciences and humanities come together. We identify our autapomorphies by comparing ourselves to our closest relatives, and we use those comparisons to understand how being human means *not* being an ape.

THE GREAT CHAIN IN CULTURAL EVOLUTION

Scientists of the 19th and early 20th centuries appreciated that the comparisons between humans and apes that ultimately helped to establish where humans fit into the scheme of nature still yielded an incomplete picture. Where, after all, did European people fit in relative not just to other kinds of animals, but to other kinds of people?

Darwin had established in biology that the fundamental mechanism of evolution is that of divergence: natural selection adapts different populations to different environments such that descendant populations are different both from one another and from their common ancestor. Therefore it was no longer apparent that evolution involved progress: species didn't become better; they became different. That naturally implied that progress was not part of the evolutionary process, as early evolutionists such as Lamarck had assumed. And thus the Great Chain of Being, which had been undermined by Linnaeus, was dispatched by Darwin as a scientific tool. As a literary device it remained, but as an explanatory or heuristic framework for practicing biologists the concept was now vacuous.

> The change from monkey to man might well seem a change for the worse to a monkey.
>
> —J. B. S. Haldane

That wasn't obviously the case when it came to examining European peoples or societies in relation to other peoples or societies. While the human species might not have clearly *risen* from the apes, as opposed to merely having diverged from them, it was fairly clear that industrialized modern Europeans *had* risen from a barbarous past—one in which people lived in villages rather than cities, had not yet learned to use metals, fought incessantly, and possibly even mated promiscuously. In other words, technology and social institutions were clearly historical developments, and were equally clearly improvements over times when those technologies and social institutions had not yet come into being. How could not having laws possibly be as good as having laws?

Thus, while there might not have been directionality to *biological* history, there certainly appeared to be a direction to *social* history. Civilization was not merely different from the barbarity out of which it had arisen; it was better. The new ethnographic literature presented Europeans with refractive mirrors of their own civilization before writing, or metals, or any of the other technological trappings they had developed. Surely, therefore, these societies represented peoples caught up in a more primitive time of history, a state out of which the Europeans had risen, but within which the natives remained ensnared.[25]

There had been other views on the matter: Jean-Jacques Rousseau, for example, had popularized the image of the noble savage, to whom civilization was not so much advancement as it was a source of corruption.[26] The idea of a Golden Age in the remote past was well-known in classical philosophy and in Christian theology. But the obvious technological inferiority of non-Western peoples left a commonsensical conclusion in the minds of Europeans. A linear sequence of Western culture history encapsulated the progress—the betterment—of society and civilization. Other societies remained frozen at stages comparable to those through which we had passed, but none had proceeded so far.

EMERGENCE OF THE MODERN CULTURE THEORY

The early part of this century saw the emergence of critical intensive ethnography ("participant-observer" fieldwork) and critical ethnology.[27] Elaborated by E. B. Tylor, "culture" took on a specialized technical meaning that involved the entire overarching edifice of our social history:

> that complex whole which includes knowledge, belief, art, morals, law, custom, and any other capabilities and habits acquired by man as a member of society.[28]

The formulation of a concept of culture permitted comparisons of various institutions across cultures, and the reconstruction of cultural evolution. Language, being one attribute of culture, had long been recognized as an evolving entity, and Sir William Jones founded the field of historical linguistics by elaborating the historical links among "Indo-European" languages before the end of the 18th century. By the turn of the 20th century, social scientists such as William Graham Sumner were deriding the "ethnocentrism" of the age, and urging the student of value systems across cultures (which Sumner called "mores") to study them without passing judgment upon them:

> Everything in the mores of a time and place must be regarded as justified with regard to that time and place. "Good" mores are those which are well adapted to the situation. "Bad" mores are those which are not so well adapted. ... This gives the standpoint for the student of the mores. All things in them come before him on the same plane. ... We do not study them in order to approve some of them and condemn others. They are all equally worthy of attention from the fact that they existed and were used. The chief object of study in them is their adjustment to interests, their relation to welfare, and their coordination in a harmonious system of the life policy. For the men of the time there are no "bad" mores. What is traditional and current is the standard of what ought to be. ... Hence our judgments of the good or evil consequences of folkways are to be kept separate from our study of the historical phenomena of them, and of their strength and the reasons for it. The judgments have their place in plans and doctrines for the future, not in a retrospect.[29]

In this program lay the basis for a conceptual revolution in the study of human behavioral variation comparable to that wrought for biological taxonomic variation by Darwin. To Franz Boas, the observable facts of cultural variation could be explained only by the unique ecological and historical circumstances of each culture. Prehistoric people in Europe had used tools made of stone, like contemporary peoples of Australia, but it was no longer clear what we could learn about the life of a "stone age" European from the life of a "stone age" Australian—for they lived in different environments and had different histories. The fundamental nature of apparent similarities would have to be established, not assumed.[30]

Further, entire cultures were sufficiently complex as to defy linear scaling. One could certainly choose an arbitrary criterion (such as technological sophistication) and classify cultures on that basis. Yet technologically *simple* cultures often had very *complex* languages and social systems. Thus, while ranking cultures in terms of technology was to some extent possible, it was based on an arbitrary criterion, and failed to represent the differences that ranking by other cultural criteria would yield.

American anthropology under Boas therefore came to adopt the position known as cultural relativism, whereby one analyzes cultures as far as possible without judging them except in the context of their own history, ecology, and belief systems.[31] This naturally undermined the possibility of discerning progress in cultural evolution—for the discernment of progress is quite simply a value judgment about the relative merit of cultures.

What, then, of the progress so apparent to earlier students of cultural evolution? It was now seen to be illusory, merely the commonsensical ethnocentric judgments of an immature science. The maturation of anthropology under Boas lay in precisely the same place as the maturation of biology under Darwin: the study of change without the framework of progress.

Culture obviously changes, but it does so by complex mechanisms. The important issue to the early 20th-century anthropologists was: Does it get "better" in any meaningful sense? If so, then is our culture "better" than that of our ancestors? And if it is, then is our culture "better" than that of the natives of Australia or the New World? During this period, these questions were all being answered now in the negative. It was not that civilization is degrading, which is also a value judgment, but simply that cultural change occurs outside an objective system of values. And because culture is a complex integrated unit, any change in one component of culture would lead to changes (usually unforeseen) in another. Boas and his students pointed to the many problems in our own culture as evidence that for each cultural problem solved, another is raised for the next generation. Ultimately we change, but in no self-evident way do we get better—except technologically, and as we of the nuclear age well know, that improvement has been as mixed a blessing as any example of culture change.[32]

Thus Boas brought cultural theory to its logical culmination in the 20th century. Darwin had undermined the biology of anthropocentrism and made it no longer possible to assert that the human species is "better" than a species of mole, for they are simply divergent offshoots of a common ancestor. So, too, Boas destroyed the underpinnings of ethnocentrism by which Western society saw itself as superior to other lifeways—it was different all right, but value judgments were ultimately based on arbitrary criteria. Western and non-Western societies were simply examples of the diverse ways of being human.

This did not mean that we are never allowed to evaluate *aspects* of other cultures; Boas himself was an outspoken critic of the social policies of Nazi Germany—the ultimate demonstration of admirably "advanced" technology in the service of a degraded system of values.[33] However, judging that one society places a greater value on human rights, and judging that society to be superior—in any kind of objective sense—are very different things. Again, by analogy to biology, if the standard of comparison is thinking, then (anthropocentrically) humans are "better" than seals, but if the standard is swimming, the seals win. Likewise, if the standard is technology, then Americans are (ethnocentrically) superior to the !Kung San of the Kalahari desert. But if the standard is the integration of the elderly into the fabric of social life, then very little self-reflection is required to appreciate that Americans are inferior. Most important, however, cross-cultural comparisons and a strong dose of humility may combine to give us an idea of how to improve our own society in order to come closer to meeting our own standards.

CHANGE WITHOUT PROGRESS: THE BIOLOGICAL AND SOCIAL HISTORY OF THE HUMAN SPECIES

All science is ultimately comparative—examining different things and explaining why they are not the same. The question of our place in nature, and establishing a scientific basis for it, has involved the maturation of sciences to the extent that such comparisons could be made value-free.

It is probably a part of general human nature to try to find meaning in events, and to the extent that history consists of events, we often try to discern overall meaning in the history of life or of society. Noting what we are, and the stages we passed through to get here, it has proved tempting to see evolution within a linear framework as culminating in our species and our society. But modern views reject the idea that we have evolved "toward" humans, or that society has evolved "toward" industrialism.

The reason is that history occurred once. Humans evolved from apes just once. Thus, we know apehood is a prerequisite for humanhood, for the only time humanhood emerged, it did so from apehood. But being an ape does not destine one's descendants to be human. We do not know precisely how humans came to be descended from some apes, but most apes did *not* have humans as descendants. In other words, being an ape is necessary, but not sufficient, for one's descendants to evolve into humans—for that is a very rare, unlikely event. It is not the destiny of apes to become human, for evolution is divergent, not linear. This is the legacy of Darwin's revolution.

Destiny, indeed, is a theme that recurs in non-Darwinian thought, for without the framework of progress imposed upon our species and all others (for example by the Jesuit evolutionary philosopher Teilhard de Chardin[34]), Darwinian evolution robs us of our ability to "predict" our biological future. The Darwinian revolution replaced a fate of progress with indeterminacy. While this may be less comforting, it is also less fatalistic: an intelligent species now has the ability to shape its own destiny. Prior to Darwin, even an intelligent species did not realize it could do more than simply live out its destiny.

Thus the divergent change that we now appreciate in nature has only short-term goals—each species tracks the environment. There does not seem to be an overarching long-term goal toward which humans did or will evolve or toward which other species or cultures are evolving. Indeed, there cannot be such a goal if evolution is mainly divergence; such goals exist only within a linear framework.

A more roundabout way of conferring unwarranted ranking on other societies or species is to consider their "potential." This, as will become apparent, is one of the major difficulties in the study of human variation, for we cannot study potential. The apes that did evolve into reflective, intelligent humans obviously had the potential to do so (because they did), but we cannot know that the apes which did *not* evolve reflection and intelligence *lacked* the potential to do so. In other words, evolutionary potential is only a retroactive concept, and can only be discerned in a small minority of cases (those that *did* are always a minor subset of those that *could have*). Therefore it is of little use as an explanatory device.

Our place in nature is an ambiguous one. We are made-over apes in a biological sense, and made-over hunter-gatherers in a social sense. However, both the substrate from which we emerged and the make-over we received are important to acknowledge in the study of our species. To ignore either one is to answer only half the question: Where did we come from?

NOTES

1. Popper (1963).
2. Medawar ([1965] 1984).
3. Quoted in Bernal ([1939] 1967:9). See also Hardy (1920, 1940).
4. Greene (1959).
5. Tyson ([1699] 1960).
6. Montagu (1943:227).
7. Singer (1957).
8. Gould (1983).
9. Lovejoy (1936).

10.   Lovejoy (1936) discusses a number of these scholars, including Leibniz, Voltaire, Bonnet, and Pope.
   11.   Eldredge (1982, 1985), Knight (1981).
   12.   The names of these orders given in the text are modern names. Linnaeus names seven other orders of mammals (Primates, Bruta, Ferae, Glires, Pecora, Belluae, and Cete), most of which are no longer in use.
   13.   Tobias (1978).
   14.   Buffon (1749–1804).
   15.   Buffon (1753), "L'Ane," in Buffon (1749–1804).
   16.   Gillispie (1951), Rudwick (1976), Bowler (1984).
   17.   Cuvier (1829).
   18.   Lamarck ([1809] 1984:20–21).
   19.   Hull (1984).
   20.   Chambers (1844), Spencer (1852), Glass, Temkin, and Straus, (1959).
   21.   Darwin (1859:420).
   22.   Goodman (1962).
   23.   Goodman et al. (1989), Marks (1991, 1992b).
   24.   Kleinschmidt and Zgouros (1987).
   25.   Hays (1958), Greene (1959), Stocking (1987).
   26.   Rousseau ([1755] 1984).
   27.   Kuper (1988).
   28.   Tylor (1871).
   29.   Sumner ([1906] 1940:65–66).
   30.   Boas (1896).
   31.   Herskovits ([1955] 1971).
   32.   Boas (1928).
   33.   Barkan (1992).
   34.   Teilhard de Chardin (1959).

2

Processes and Patterns in the Evolutionary History of Our Species

The processes of evolution operate on gene pools, and accrue to species. The processes of evolution are consequently genetic processes, but the patterns they produce are taxonomic. Relating them to each other in practice involves the creative use of narrative.

NARRATIVE AS A SCIENTIFIC MEDIUM

All human intellectual endeavors use language: it is one of the primary autapomorphies of our species. Science is such an endeavor, and science is constructed fundamentally of language. Finding out about the universe is the main goal of science, but our comprehension of it is constrained by our own mental linguistic processes; and learning about the universe is useful only to the extent that insights can be communicated to others, which further constrains the scientific process through language. Sometimes the linguistic structure of science is obvious, as in the metaphor that attributes "charm" to subatomic particles. It can also be more subtle, as in the inference that the scientific endeavor proceeds by a sequence of (1) background information, (2) materials and methods, (3) results, (4) discussion, and (5) conclusions—simply because scientific *papers* are written that way.[1]

Though all science has special linguistic features of narrative, the science of human origins has a particularly self-conscious streak. Consequently, the structure that linguistic forms have imposed upon biological anthropology—the way in which the medium becomes mixed with the message—has been more intensively examined here than in other sciences.

How do the data become conflated with the manner of their transmission? Misia Landau draws attention to the origin of human adaptations, and how discussions of their origin have tended to be formulated

in terms similar to those found in folktales. The humble hero is challenged, grapples with his obstacle, ultimately triumphs, and garners a
reward.

> Like many myths, the story of human evolution often begins in a state of
> equilibrium ... , where we find the hero leading a relatively safe and
> untroubled existence, usually in the trees. Though he is a nonhuman pri-
> mate, he is somehow different. ... Often he is smaller or weaker than
> other animals. Either by compulsion or choice, the hero is eventually dis-
> lodged from his home. ... Having departed, the hero must move in a new
> realm where he must survive a series of tests. ... Whether imposed by
> harsh climate or by predators or other competitors, these tests are
> designed to bring out the human in the hero. ... [A] hidden figure or
> donor provides the hero with the means to overcome his enemy or attain
> his desired object. ... [E]volutionary principles operate as hidden agents
> in stories of human evolution. ... These forces bestow on the hero the
> gifts—intelligence, tools, a moral sense—that transform him into a primi-
> tive human.[2]

Certainly the popular scenarios of human evolution bear this out:
recall the humble apish hominids in *2001: A Space Odyssey*. Threatened
by other groups, one thoughtful ancestor begins to use long bones to
bash other bones; soon he begins to bash his enemies' skulls with the
new tool, and in hardly any time at all, his descendants are building
space stations.

What differs among various scientific theories is how the bits of data
are embedded into the story. And specifically in human evolution, we
are compelled to ask: Given that the process of becoming human
involved modifications of the teeth, pelvis, and brains, in what sequence
did they occur? Was the initial factor that led an ape toward (what we
now recognize as) humanity a dietary crisis, which would imply teeth
leading the way? Was it instead a forest crisis, forcing animals out of the
trees, which would imply pelves leading the way? Was it tools and the
ability to use and refine them that led the way? Or was it something
else, perhaps no crisis at all? In other words, how do we connect the
dots provided by the data?

ADAPTATION STORIES

The fit between how an animal appears, what it does, and where it
lives was known to Aristotle, and explaining that fit has been a major
focus of the study of life for centuries.[3] Certainly the most enduring
explanation is the one holding that each species has no history, having

existed from the beginning doing just what it does, looking as it does, living where it does.[4] The last influential version of this theory was the "watchmaker" analogy of William Paley.[5] Here, the intricate match of the organism's parts to one another, to its lifeways, and to its surroundings is likened to the construction of a precision timepiece. As timepieces are designed and assembled by ingenious and kindly old men, Paley reasoned, so are species—designed and precision-crafted to fit perfectly into their space in nature.

Darwin, in the *Origin of Species*, suggested an alternative that modern science has found more fruitful: that each species has a history. The fit of an organism to its environment, therefore, must be the result of a long process of *adapting*. How does this process take place? By the fact that all members of a species do not survive and reproduce with equal efficiency. The consistent ability of organisms with certain attributes to perpetuate themselves more efficiently than other members of the same species lacking those attributes results in the apparent transformation of a species through time. That is certainly the origin of the various breeds of pigeons and dogs that have been bred into existence during human history.

It is the environment, reasoned Darwin, that determines which characteristics permit an organism to survive and reproduce disproportionately in nature. Where the changes in animal breeds are wrought by human intervention, there is a process of selection by the breeders by which animals with certain attributes are actively and consciously chosen to produce the next generation. By analogy, in the natural world, nature does the selecting—passively and unconsciously, and thereby requiring many generations, but the net result is the same: descendant populations come to diverge anatomically from their ancestors, and from other descendant populations.

Anatomical divergence is not random, however. Selection by animal breeders produces animals that conform to a particular model the breeder has in mind—whether that model is a population of large Great Danes, a population of small chihuahuas, lithe and sleek greyhounds, or cuddly bassets. The selective action of nature, however, favors simply the survival and propagation of those organisms that are best equipped to function within that particular environment.

The fit of an organism to its environment, therefore, is the result of its history, the action of natural selection gradually modifying populations over the course of time. This explanation for the existence of adaptations, Darwin's explanation, immediately found wide favor within the scientific community, and remains the accepted process by which adaptation is achieved by species. On the other hand, adaptation to the environment is not the only evolutionary force operating. Sometimes, for example, an adaptive change in one part of an organism physiologically necessitates a change in another part of the organism. The second change, therefore, would not itself be an adaptation attributable to the environment, but would require an understanding of the processes by which bodies grow and develop. An increase in body size, for example, requires a larger increase in some physiological systems to maintain function, simply by virtue of the fact that their efficiency is determined by the area of an organ, but area and volume are not linearly proportional to each other.[6] Alternatively, sometimes morphologies vary without tracking the environment. Is there, for example, an environmental necessity that could account for the existence of a pug nose rather than a flat, aquiline, or hooked nose?

Thus, the difficulty with adaptation and natural selection comes in reasoning backward: Given an adapted population, what properties are in fact adaptations (i.e., the result of natural selection, as opposed to other evolutionary forces)? Further, given a specific complex morphological adaptation, what particular aspects were favoring the survival and propagation of its bearers? In humans, the classic example is bipedalism. Given that it did evolve as an adaptation by natural selection, what is it about bipedal locomotion that dictated the disproportionate survival and propagation of bipedal apes as opposed to others? Was it greater endurance? (It certainly wasn't greater velocity!) Was it an enhanced ability to see over tall grass? (How much tall grass could there have been?) Was it an ability to use the forelimbs in creative ways? (But what about the lag time—over a million years—between becoming bipedal and use of the earliest stone tools?)

Again, narratives function to fill in the gaps, and though periodically criticized,[7] adaptive scenarios are among the most pervasive and enduring stories in science. One is virtually forced, by simply knowing how natural selection works, to try to imagine what advantage to an earlier creature a contemporary observable feature may have brought. To the extent that such a reconstructive enterprise must rely more on imagination than on experimental testing, it is in great part a narrative endeavor.

DISTURBING THE CONSERVATIVE NATURE OF HEREDITY

Evolution has happened, and organisms are adapted to their environment—regardless of how extensively our data must be augmented with stories to flesh them out and permit them to take the form of explanations for how we came to be as we are.

The processes of change in species through time are a consequence of the procreative activities of the organisms composing the species. Procreation is, in an important sense, a genetic paradox in whose nature lie the secrets of evolution. On the one hand, heredity is conservative: like begets like. On the other hand, heredity is not clonal: we are not identical to our parents. The same paradox exists above the level of the organism: a population in a single generation is a genetic copy of the generation that preceded it, but not a perfect copy. It is this tension between faithful reproduction and not-too-faithful reproduction that lies at the heart of the evolutionary process.

REPRODUCTION OF ORGANISMS: MEIOSIS

The hundreds of thousands of genes, the units of heredity, are composed of long strands of DNA. This hereditary material is organized into a manageable number of units we call chromosomes. Chromosomes are most commonly visualized in their condensed form, during cell division. At other times, when the cell is performing a function, rather than dividing, the genetic material is diffuse, spread throughout the cell nucleus. In an ordinary cell, there are two copies of each piece of DNA: one inherited from father, and one from mother. In more concrete terms, one inherits a single set of chromosomes from each parent, and a normal cell consists of two of each chromosome.

Sexual reproduction consists of a cycle of halving (*meiosis*) and doubling (*fertilization*) of the genetic material. The net result, obviously, is that the amount of genetic information remains constant from parent to child; however, there are three ways in which fidelity of transmission is undermined in the process of halving and doubling.

The process of meiosis, cell division in the sex organs, proceeds somewhat differently from the normal manner of cell division in the rest of the body, the process of mitosis (Figure 2.1). In mitosis, a cell fissions to produce another cell genetically identical to itself, and is the cell division that constitutes the growth of an organism. Rather than producing two identical cells from one, *meiosis* produces four cells with half the amount of DNA, all different from one another.

Meiosis is by its nature a faithful cell division. The DNA duplicates, as it does before ordinary cell division; and it begins to condense into chromosomes, as in ordinary cell division. At this point, since the DNA has faithfully duplicated itself, there are actually four copies of each piece of heredity, two opposite each other, called sister chromatids of the same chromosome.

In mitosis, the sister chromatids simply split apart, and one chromatid from each chromosome migrates to one or the other end of the cell. Each human cell has 46 chromosomes, and both poles of the dividing cell

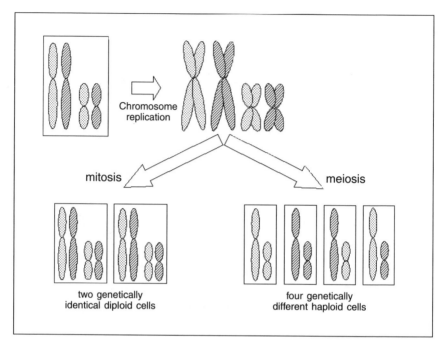

mitosis

meiosis

Chromosome
replication

two genetically
identical diploid cells

four genetically
different haploid cells

Figure 2.1. Mitosis produces two cells genetically identical to each other and
the original cell, while meiosis produces four cells with half the amount of
genetic information.

receive 46 chromatids, the full complement of genetic material. At this
point, the chromatids are regarded as chromosomes themselves. In
meiosis, however, another kind of division precedes the separation of
the sister chromatids. By a process not understood, pairs of chromosomes (for example, the chromosome #12 of paternal and of maternal
origin) recognize each other and form an intimate association. The
process is called *synapsis*, and results (in humans) in 23 visible structures
called *tetrads*, each of which is composed of two homologous chromosomes containing two sister chromatids apiece. The unique aspect of
meiosis involves the separation of homologous chromosomes from one
another, migrating to different poles, such that each pole receives one
chromosome of each of the 23 pairs, each still composed of two sister
chromatids. Homologous chromosomes separate at the first meiotic division, and then sister chromatids divide in a second meiotic division similar to mitosis.

Three processes occur that not only produce diversity, but actually
ensure extensive scrambling of the genetic material. These are: crossing-
over during meiosis, independent assortment of chromosomes during

meiosis, and the random union of gametes during fertilization. To appreciate them, it is important to note that while sister chromatids are absolutely identical (except for rare errors in copying the DNA), homologous chromosomes are not identical. Homologous chromosomes come from different parents. Each has, at any specific site (or locus) the same gene (for example, the gene for blood type) located at the tip of chromosome 9. Yet the particular information residing at the location may often differ across homologous chromosomes—one could have the information encoding blood type A, blood type B, or blood type O. The different genetic variants—different answers to the same hereditary question—are called alleles.

Since father and mother are genetically different from one another, the two homologous chromosomes that originated from them must have different alleles on them. Indeed, since there are probably tens of thousands of genes on an ordinary chromosome, one chromosome #7 may well consist of a long string of different alleles from the other chromosome #7, its homolog. Those strings of alleles could conceivably be passed on intact indefinitely. However, meiosis actually precludes that. During synapsis, the intimate pairing of the homologs, a process known as *crossing-over* takes the alleles on part of one homolog and attaches them to the alleles of a different part of the other homolog (Figure 2.2). The resulting recombinant chromosome contains all the right genes, but is now a mixture of DNA that came from the paternal and maternal homologs. Thus, one cannot pass on intact a chromosome received from one's parent.[8] Meiosis ensures a scrambling of the maternal and paternal alleles of every chromosome each generation.

The cell also begins with an entire maternal and paternal complement of chromosomes, yet ends with only a single complement. Conceivably the chromosomes inherited from one parent could travel together

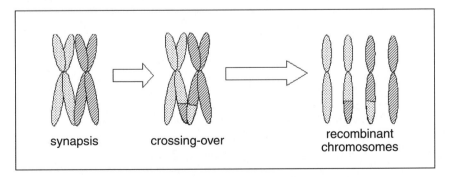

synapsis   crossing-over   recombinant chromosomes

Figure 2.2. Crossing-over combines previously unlinked alleles on the same chromosome.

through the germ lines of the generations indefinitely, but again the process of meiosis precludes this. When the homologous chromosomes separate—23 going one way, and 23 going the other way—the 23 that travel together are a motley assortment of maternally and paternally derived chromosomes. Indeed, this was the key insight of Gregor Mendel in 1856 that laid the foundations for modern genetics when it was finally appreciated at the turn of the century. Each cell must get a chromosome #1, a chromosome #2, a chromosome #3, etc., but that assortment of chromosomes will be, on the average, half paternally-derived and half maternally-derived. Thus, the process of meiosis guarantees that not only combinations of alleles, but also combinations of chromosomes are fully scrambled each generation.

Finally, the restoration of two chromosome sets at fertilization, which involves the union of gametes from different organisms, is also a process that generates diversity. Conceivably, sperm bearing certain chromosomes from one parent could always carry out the fertilizing of the egg. Instead, however, the union of gametes is random: a sperm with any genetic constitution from the father has an equal opportunity to fertilize an egg with any genetic constitution from the mother. Again, this maximizes the amount of genetic diversity that the union of gametes can yield.

Thus, while the complementary mechanisms of meiosis and fertilization serve to pass on the hereditary instructions from parents to offspring, they also shuffle the

> Rarely are sons similar to their fathers: Most are worse, and a few are better than their fathers.
>
> —Homer, *The Odyssey*

contributions. While any particular instruction is passed on generally intact, every generation finds the instructions in new combinations that were not present in the parents. The process of reproduction in organisms embodies a tension, then, between faithful transmission across generations and the generation of genetic novelty. This tension exists as well at a level "above" the organism.

REPRODUCTION OF POPULATIONS: THE GENE POOL

One of the major advancements of 20th-century science was the development of population genetics, which gave a quantitative framework to Darwin's theories.[9] Mendel had inferred that all genes are present in two copies in organismal bodies. The genetic constitution of an animal is called its genotype, and the observable characteristic that results from the genes is called a *phenotype*. The relationship between genotype and phenotype, or how traits of organisms come to be expressed through the growth and development process, from a single fertilized egg with lots of functioning genes, is still a central problem in genetics. Mendel inferred also that if an organism has two different alleles of a gene, a condition called *heterozygosity*, sometimes only one of the alleles will be expressed as a phenotype. Thus, a person with genotype AA has the same blood type as someone with genotype AO: both have the phenotype of type A blood. The phenotype that is expressed in the heterozygote (in this case AO; the person with the AA genotype is a homozygote) is said to be dominant; the one that is masked is recessive.

Not only organisms reproduce themselves; populations do as well. Any population with genetic diversity has a unique genetic composition—say 10 percent of one allele and 90 percent of another for one gene system, or 30:30:40 for another gene system with three alleles. The genetic composition of the population, which transcends the specific genotypes of the specific organisms that compose it, is called the gene pool. It is a theoretical summation of all the gametes in a population. In this formulation, organisms are regarded as simply transient and short-lived packages of heredity; they do the business of ensuring that the genes get transmitted, but are themselves ephemeral. The gene pool thus has greater breadth than any individual organism, and greater longevity (since it endures many generations), and in evolutionary genetics it is the object of analysis.

Shortly after the turn of the century, Hardy and Weinberg showed mathematically that a gene pool of a population is transmitted faithfully by its organisms under the rules of Mendelian inheritance. This has come to be known as the Hardy-Weinberg law, and demonstrates that in a population with a given amount of genetic diversity, the same diversity will be found in the next generation—indeed, will be perpetuated indefinitely.[10] This law is thus an equilibrium statement, a description of like begetting like at the level of the population or gene pool.

Once again, there is a fundamental tension between the population reproducing itself faithfully, and the manner by which the next generation is not its perfect replica. In this case, the foundations of evolutionary genetics lie in analyzing the ways in which the Hardy-Weinberg equilibrium can be perturbed. By the 1930s many geneticists appreciated that there are four basic ways in which the Hardy-Weinberg equilibrium describing the perfect genetic replication of a population in the next generation can be violated: these are the modes of microevolution.[11]

MICROEVOLUTIONARY PROCESSES

Mutation is the ultimate source of new genetic variation. All alleles begin as mutations within single organisms. By itself, any particular mutation does not arise often enough to be significant in the gene pool, since mutations are by their nature rare. They result from a mistake—a slightly unfaithful copy of the DNA—prior to cell division, and can enter the next generation if they happen to be in a gamete that participates in fertilization. Every generation every organism, in all likelihood, has a complement of brand-new mutations. Though any specific mutation is rare, there is much DNA that needs to be copied, and there are many organisms undergoing DNA copying. Therefore, a constant infusion of new genetic variation exists in any population.

Most mutations are not expressed as phenotypes, or are expressed in so subtle a manner as not to affect the overall quality of life of the organism. Chapter 8 shows that this expression is due to the fact that most DNA is non-coding, and can tolerate a considerable amount of diversity while still maintaining the integrity of its function. Most mutations that have a direct and discernible effect on the phenotype are injurious. The instructions for the construction of a functioning organism are indeed intricate, and it is certainly much easier to foul them up with a random mutation than to make them better or more efficient.

Understanding mutations is where Paley's watchmaker analogy is still useful, though the watch (i.e., the genetic blueprint) is now taken to be formed by natural agencies over the eons.[12] Making random adjustments to the watch is unlikely to make it better. On the other hand, the incorporation of certain adjustments has helped, for watches today aren't what they were in 1802. The important difference to bear in mind is that the history of timekeeping involves a conscious activity to improve the watch, so the changes were made with a goal in mind, while the mutations at the heart of microevolution occur mindlessly and perpetually. Their incorporation into the genetic machinery, into the formation of new characteristics, is again that central problem—the generation of phenotypes from genotypes.

The introduction of a new genetic variant can occur either by a new mutation, or through the entry into the gene pool of a mutation that arose in another population. In humans, the most common form of the latter process is intermarriage, and the genetic consequence is known as gene flow.

Nevertheless, a new genetic variant in a single organism is by itself a singularly minuscule evolutionary event. To have significance, a newly-arising or newly-introduced mutation must spread over the generations. It does this in two fundamentally different ways: by natural selection and by genetic drift. In the former case, the result is adaptive change in the population; in the latter case, the result is non-adaptive change.

The genetic translation of Darwin's principle of natural selection involves organisms with different genotypes consistently reproducing more efficiently than those with other genotypes. The result is a gene pool disproportionately represented by the prolific genotype. If the most efficient reproducer is a homozygote, selection is directional (toward replacement of one allele by another); the gene pool comes to be characterized by the particular allele possessed by those homozygotes, and the other allele is supplanted. If the most efficient reproducer is a heterozygote, then a balance is achieved in the gene pool between the frequencies of the two alleles. Either way, the population comes to be better adapted genetically to its local environment by virtue of the transformation of its gene pool.

Natural selection, therefore, translates genetically into precisely Darwin's conception: nature's watchmaker, the force that adapts organisms. By the consistent out-reproduction of organisms with certain genotypes over those with other genotypes, the gene pool comes to be altered over the generation in the image of the reproducers. The difference between Darwin's version of natural selection and the one we have just articulated is that Darwin focused on phenotypes. But phenotypes are important in evolution only insofar as they may be the manifestations of underlying genotypes. A phenotype that is the result of a strictly environmental agent, such as consistent taillessness in mice due to consistent amputation, is not evolutionary: it does not affect the gene pool. The mice of the 21st generation would be born with tails as long as those of the first generation, for their genetic constitutions were not causing the phenotype. Indeed, this was one of the classic demonstrations, by August Weismann, that environmental agencies do not have a direct effect upon the genetic constitution.[13] Natural selection, therefore, "selects" among phenotypes—but its evolutionary importance is how it affects the composition of the gene pool.

The fourth manner in which gene pools change is the most abstract, and also the most under-appreciated. It is due to those random deviations of life from mathematical predictability, and is called genetic drift. The Hardy-Weinberg law implies that a gene pool consisting of 1 percent of a certain allele will generate descendant gene pools with 1 percent of that same allele, assuming no new variation arises (no mutation), no new additions are made (no gene flow), and all individuals reproduce at equal rates (no selection). But consider a population that consists of relatively few organisms. Suppose a population of 50 reproduces, but the next generation has only 20 organisms. An allele at a frequency of 1 percent in the parent population would be mathematically troublesome. There are 100 total alleles in the parent population, since each organism has two alleles; therefore there is one copy of the allele in question, and 99 of the other. In the descendant population, however, if there is just one copy, the frequency is now at 2.5 percent (since there are only 40 total alleles); if there are no copies, it is at 0 percent. This means that simply by manipulating the number of individuals in a population, microevolution has occurred, for the composition of the gene pool has changed.

Genetic drift is the property of finitely sized populations to deviate from the mathematical expectations. The magnitude of the deviation is inversely proportional to the size of the population (in the above example, one copy of the allele in a population of just five organisms would mean the allele has risen in frequency from 1 to 10 percent). If a population were infinitely large, genetic drift would not operate; in the real world, however, it may be a very significant factor. The changes brought about by genetic drift do not track the environment; indeed, they are simply random fluctuations in the gene pool, and so are non-adaptive.

There are three main ways in which genetic drift can operate. The first is founder effect, the origin of a new gene pool from a subsegment of a parent population. If the founder's gene pool is much smaller than that of the parent population, then it may not be fully genetically representative of it. Therefore, even if the descendant population expands and comes to achieve large numbers, it may not be identical to the parent population because its founders were not a full sample of that population. The second manner is in the spread of "neutral" alleles: those that have no impact on the ability of its bearers to survive or reproduce more efficiently. Any particular neutral allele is very unlikely to spread much, since it begins in but a single organism. But if there are many different examples of these alleles, constantly arising in the gene pool, then some will spread to large frequencies randomly. It's like winning the lottery: the chance of any particular individual doing it are very small, but there is a constant supply of winners drawn from the numerous players. Finally, the third manner by which genetic drift acts is in population crashes, catastrophic cutbacks in the size of the gene pool. In this case, the descendant population is in effect a founder population, and the same uncertainty as to its representativeness applies.

Genetic drift acts to deviate populations genetically in random (non-adaptive) ways. Natural selection acts to diverge populations in adaptive ways, tracking what the environment demands. Gene flow homogenizes populations, and mutation has little direct effect on their gene pool by itself. Each of these microevolutionary forces is a violation of the Hardy-Weinberg equilibrium, a way in which the stability of inheritance is undermined.

A fifth violation of Hardy-Weinberg does not affect the composition of the gene pool directly, but instead affects the distribution of genes into genotypes. A population of 50 people that is 50 percent allele A and 50 percent allele O could be composed of 25 AA homozygotes and 25 OO homozygotes, or (most extreme among many options) 50 AO heterozygotes. In one sense, the apportionment of the genes into genotypic packages does not matter for microevolution, since the gene pool is identical in composition in both cases. On the other hand, the phenotypic distribution is considerably different in both cases: there are many individuals with type O blood in the first case, and none in the second. If natural selection, which operates upon phenotypic diversity, is operating on these populations, then the distribution of alleles may work in synergism with the selective force to alter the gene pool. In other words, there is phenotypic diversity in the first population, but not in the second. Natural selection cannot act on the second population, since all the individuals are identical in this characteristic, but it can work on the first population, since individuals with different phenotypes represented.

Thus, the distribution of genes into genotypes may augment the change of the gene pool, while not itself changing the gene pool. The fifth violation of Hardy-Weinberg is one that affects the distribution of genes without changing the gene pool. This violation is inbreeding, the mating of relatives with one another. Inbreeding has the effect of increasing homozygosity and reducing heterozygosity in a population. In small populations, it may be impossible to choose a mate not related to you in some way; this is why many isolated populations have elevated frequencies of alleles that are rare in other, more cosmopolitan populations. Thus, inbreeding can work in synergy with founder effect (genetic drift) to alter gene frequencies.

MACROEVOLUTIONARY PROCESSES

Not only do gene pools of populations diverge from one another, but the organisms composing them often develop aversions to mating or incapacities to mating across populations. In other words, not only are the gene pools different, but they are effectively sealed off from one another. We recognize this as the formation of a new species, or speciation. The formation of new species marks the break between microevolution and macroevolution. Processes that occur below the level of the species are microevolutionary; those above the species are macroevolutionary.

Speciation is the ultimate source of taxonomic diversity in the world. The reason that species differ from one another is two-fold. First, their gene pools are different. And second, their gene pools *cannot* become more similar to one another by interbreeding (gene flow). What is important about these dual processes is that they *are* dual processes—consequently, there is no way to predict for certain that because the gene pools are different by a certain amount (or the phenotypes are different by a certain amount), the two populations are consequently unable to interbreed. Apparently the ability to interbreed is not closely related to the acquisition of genetic or anatomical divergence. This, of course, makes the discrimination of species particularly difficult when analyzing fossil material.

Extinction is the opposite of speciation, the elimination of taxa from the living world. There are essentially two kinds of extinction: background extinctions and mass extinctions, each with a radically different implication for interpreting patterns in the history of life. Background extinctions are those events that mark the end of a species' attempts to compete successfully and to thrive—it represents failure on the part of the entire species in the struggle for existence. Mass extinctions, by contrast, are the results of major ecological catastrophes. Whatever the cause, the effect is a termination, in a relatively brief geological span of time, of many different kinds of species. These species haven't failed in any ordinary kind of competitive struggle—they simply lost a major lottery in life. The result is an ecological vacuum and an evolutionary free-for-all for the surviving species to expand. This may well be the way in which the mammals came to supplant the dinosaurs 65 million years ago.

EVOLUTIONARY NARRATIVES

History is related in prose, and evolutionary history uses a particular form of prose, derived from the processes inferred to be at work. At their most fundamental, histories of the human species focus on the competitive edge of becoming human as they emphasize the role of natural selection in that history. Becoming human thus becomes the story of developing an edge on the apes, becoming favored in the eyes of nature.

The luck of the draw is emphasized with genetic drift, and the precariousness and contingency of our existence is the moral of the extinction of the other hominids. Gene flow was a major mechanism invoked by racist geneticists in the 1920s, whose narratives focused on the rise and fall of civilizations through the purity of their gene pools. For example, the influential author Madison Grant maintained in *The Passing of the Great Race* (1916) that art, law, and morals were perpetuated through purity of blood—civilizations fell ultimately through gene flow.

Natural selection is deterministic in that it "pulls" a species in a specific direction toward whatever is advantageous in the particular environment. Genetic drift is random in that it "pulls" a species in no particular direction consistently toward or away from what might be advantageous. The deterministic quality of natural selection involves a highly localized goal—better adaptation. Adaptation is to immediate circumstances and surroundings, yet we often imagine goals to be long-term and transcendent. This difference sometimes gives natural selection an additional literary property—teleology—that is not merited.

The problem surfaces when we try to explain apparent evolutionary trends. For example, given an early species with a small brain, a late species with a large brain, and a temporal intermediate with a medium-sized brain—do we infer that there was consistent selective pressure for brain expansion? If so, doesn't that imply that the "early" and "middle" species (call them *Homo habilis*, *Homo erectus*, and *Homo sapiens*) were simply precursors, on a path to becoming human?

And yet, *Homo erectus* existed for well over 1 million years, about three times as long as *Homo sapiens* has existed. It is thus very paradoxical to assign an obviously quite successful species the status of "precursor-on-the-path." That assignment is an expression of chauvinism—the judgment that at any point in time there are two kinds of creatures: those successfully on their way to becoming human, and those that are not.

In fact, there are no creatures at any time "on their way" to becoming human, any more than we are "on our way" to becoming whatever some of our descendants will be in 2 million years. Though there is only one past, there are many possible futures: how then can we say we are "on the road" to any one of them? Knowing that there is only one past, it is tempting to see our ancestors as imperfect, partly formed, a way station between apes and humans. This thinking implies a single transcendent path, when we also know that evolution has involved branching—forks in the road. For example, *Homo habilis* had as descendants Neanderthals, Leonardo da Vinci, and Charles Manson—and it is no more correct to consider *Homo habilis* as on its way to becoming Leonardo da Vinci than on its way to becoming either of the others. This is the destiny fallacy mentioned in the previous chapter: We cannot infer that it was the destiny of *Homo habilis* to produce Leonardo da Vinci simply because *Homo habilis did* produce Leonardo. It produced many others in addition, and had history been only very slightly different, it might not have produced Leonardo at all.

Thus, although we can write about the process of becoming human (some even use the term "hominization"), such a process never really existed. There was simply a set of processes and events, one result of which was the human species.

Nevertheless, when we look at the diverse creatures that share a special relationship to humans, but not to apes, we face the strong temptation to categorize them as being either our precursors or not our precursors. One manifestation of this temptation is the allocation of tools to fossil taxa. Tools imply the existence of an intelligent creature who made them, but if more than one hominid species is present, how do you write the story? In general, the story has been that the species most like us—and only that species—made the tools. Thus when the Leakeys discovered Zinjanthropus (now generally called *Paranthropus boisei*) at Olduvai Gorge in 1959, they naturally allocated the stone tools to that species as well. A few years later, when *Homo habilis* was found there, the creation of the tools was transferred to *Homo habilis*. In fact, tools could probably have been made by *Paranthropus*, though it is not in our direct ancestry.[14]

HUMAN MACROEVOLUTION

The three major detectable differences between humans and apes are in the mode of locomotion (bipedalism), teeth (small canines, with an emphasis on the rear teeth), and general means of survival (culture). Other differences between humans and apes exist, physiologically and socially, but these three are the ones that are preservable in the material record.

In the late Miocene, 10 to 6 million years ago, the primate fauna was profoundly different from the contemporary primates. Whereas the modern large ape clade com-

> Nevertheless, it is even harder for the average ape to believe that he has descended from man.
>
> —H. L. Mencken

prises only four genera (*Homo*, *Pan*, *Gorilla*, and *Pongo*), considerably greater diversity was represented in the Miocene clade, which included *Sivapithecus*, *Dryopithecus*, *Ouranopithecus*, *Graecopithecus*, *Gigantopithecus*, *Lufengpithecus*, *Oreopithecus*, and possibly many others.[15]

The descendants of one of these—or perhaps of a genus still unknown—adopted the bipedal gait. With the adaptation of their grasping big toe for weight-bearing, they became less skillful climbers and clamberers in the forest, and took instead to the savanna. The earliest evidence of bipedalism consists of footprints preserved in volcanic ash at Laetoli, Tanzania, about 3.7 million years ago. Skeletal remains of primates adapted to the bipedal habit are somewhat younger than the footprints, slightly more than 3 million years old, and are called *Australopithecus afarensis*. While bipedal, this species retained many primitive features of the apes: larger canines, a small brain, long arms, and strong shoulder muscles.[16]

The descendants of *Australopithecus afarensis* were of three kinds, which, originally diverging from other apes in their locomotion, now diverged from them dentally as well. Teeth that are diagnosably different from those of African apes can be identified as far back as 4.5 million years ago—but they are still more similar overall to those of apes than to those of humans.[17] Two to 3 million years ago, however, *Australopithecus africanus* (first represented in fossil collections by the Taung child's skull discovered in 1924 in South Africa) existed without the large canine teeth characteristic of catarrhine, especially male, primates.[18]

Paranthropus, with enlarged chewing muscles mobilizing great molar teeth set in immense jaws, nevertheless had front teeth (including canines) smaller than our own. In other words, whatever tendencies the human line had to reduce the canine teeth and thereby differentiate themselves from the apes, were taken even further in the hominid lineage by *Paranthropus* (Figure 2.3).[19]

Thus, we attribute to *A. afarensis* the bipedal innovation of the hominid group, and to the later australopithecines the dental innovation of the hominid group. The third descendant of *A. afarensis* may have been a more recent descendant of an early *A. africanus*, and is identifiable more than 2 million years ago, the genus *Homo*.[20] It is to *Homo* that we attribute the origin of the reliance on cultural adaptations that now characterizes our own species.

Though apes were cosmopolitan in their distribution across the Old World of the Miocene, all the hominid genera originated in Africa. By about 1 million years ago, however, all hominid species save one had become extinct; the survivor was *Homo erectus*, first known in Africa between 1.5 and 2 million years ago, and branching out into Asia about

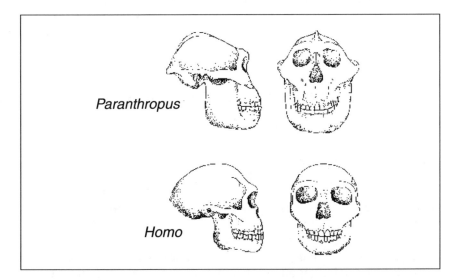

Figure 2.3. Two hominid genera, contemporaneous about 1.5 million years ago. Above, *Paranthropus*; below, *Homo*.

1 million years ago. With the beginnings of this species may well have come the limb proportions (short arms, long legs) that characterize modern humans.[21]

A recent controversy centering on evolution through genetic drift or gene flow involves the origin of the human species from *Homo erectus*. In one version, focusing on the similarities between people inhabiting the same place at different times, gene flow operated to homogenize *Homo erectus* to the extent that local populations of that species were all able to evolve into local populations of *Homo sapiens*.[22] An alternative view, emphasizing the similarity of people in different places at the same time, is that only a small founder population of *Homo erectus* actually evolved into *Homo sapiens*, such that local populations of modern humans are *not* descended from local populations of *Homo erectus*, but rather from a single common population of *Homo erectus*.[23] The contrasting narratives can be read either in terms of the caprice of evolution and the vagaries of the founder effect in producing the human species through a series of population bottlenecks and migrations, or in terms of long-term geographic stability.

The founder-effect model is currently favored, which represents a shift away from explanations for human diversity that invoke long-term stability of "racial type." The implication of the bottleneck-colonization model for human variation is that whatever similarities contemporary humans have to the inhabitants of the same place in the middle Pleistocene are due either to convergent adaptations or to a small degree of genetic continuity—but those similarities are dwarfed by the resemblances of modern humans to one another and specifically to Africans of the middle Pleistocene.

LINKING DATA INTO HISTORIES

The reconstruction of history into a scientific narrative involves the creative linkage of data. Certainly the bits of data at our disposal are related in some significant way; it is in this linkage of the facts that narratives of human evolution tend to diverge from one another. Bipedalism, for example, "freed the hands" from locomotion for tool use. However, because the earliest evidence of stone tools is over 1 million years later than the evidence of bipedalism, the relationship between the two activities is not perfectly clear.

Further, many uniquely human attributes are related in some way to detectable skeletal autapomorphies, but cannot themselves be inferred. The shape of a bipedally adapted pelvis and the size of a hominid baby's head place constraints on the human childbirth process, which requires human infants to rotate their heads while in the birth canal. Some unique aspects of the hominid birth process can be inferred for australopithecines from their pelves. But humans do not consume the placenta, as other primates do.[24] When did our ancestors stop? Here is a division between humans and apes, but how can it be dated or integrated into the fabric of evolutionary history?

Many interesting questions cannot be approached by normal scientific means, simply because the relevant information is not preserved for us in fossils. What explains the loss of body hair in humans? What led to the emergence of sexual dimorphism in the facial hair and sparse body hair of humans? What of the retention of hair only in the smelliest parts of the body (pubic and axillary areas) at puberty, and on the head at all ages? What of the multiplication of sweat glands in the skin, which enables humans to dissipate heat through evaporation rather than by panting?

When and why did the human diet change from one in which meat contributes a negligible nutritional supplement (as in the other apes), to one in which it figures prominently? Was the infusion of dietary meat the result of males going off to the hunt, or of males and females scavenging carcasses together?

How and when did human communication adopt the symbolic vocal system we now recognize as language? Is its expression identifiable as the expansion of the cranium, or much later as cave art? When did speech emerge as an audible expression of the mental processes of language? When did humans begin to cry to express their sadness, augmenting the sobbing of unhappy ape ancestors with flowing tears of unhappy humans?

When did the patterns of sexual dimorphism characteristic of humans and not of other primates emerge? We can see dimorphism in canine teeth, similar to other primates, in *Australopithecus afarensis*, but rarely in subsequent hominids. But in which taxa did the body composition of females begin to change at puberty, resulting in the patterns of fat deposition characteristic of modern human women? How did this relate to the concealment of ovulation, such that humans become sexually receptive not simply when they are fertile, as generally in other primates, but when they are not fertile as well? Or to loss of the small bone in the tip of the penis, known as the baculum, concomitant with an enlargement of the organ itself? Does this different pattern of sexual dimorphism reflect a different pattern of competition for mates in hominid prehistory?

Finally, when and in what taxa did the social patterns characteristic of humans begin to emerge? In particular, when did males and females begin to take on different responsibilities for the welfare, and particularly the nutrition, of the group? How did eating become a social activity, at a home base rather than on the fly, with rules for the division of food and for its preparation? When did marriage, representing the legitimization of birth, establishing reciprocal obligations between families, and creating a stronger bond between *father* and child than in the great apes, come to be?

PATTERNS IN THE EVOLUTION OF SPECIES AND CULTURE

Perhaps the most striking feature of human evolution is the niche that characterizes the species: symbolic creative thought and its expression in ways of coping with the environment and obtaining the necessities of life through the use of material culture.

This niche probably originated with the earliest bipedal hominids, as our ape cousins are now known to use tools in several contexts. The hominid difference seems to involve a full genetic commitment to nongenetic adaptations. Other species became bipedal—for example, kangaroos and tyrannosauruses—so bipedalism doesn't seem to be a sufficient condition for becoming a cultural being. Perhaps symbolic creative thought was the effect of bipedalism on a primate, an already visually oriented and tactile creature.

The significance of culture is that it provides a different mode of adapting, one that is easily malleable in direct response to local conditions. Because culture is symbolic, it provides more flexibility than learned behaviors in other species. Because culture is communicated socially rather than genetically, it can spread faster and thereby affect more organisms more directly and more rapidly than other kinds of adaptations. And finally, because culture is cumulative, it can develop its own history largely independently of the biological history of the organisms that utilize it. For example, the DC-10 is a descendant of the biplane—though there may be no clear biological relationship between the designers of the biplane and the designers of the DC-10.

The interactions between culture and human biology are complex, and will be explored at length in later chapters. Possibly the most interesting relationship between them is the apparently causal effect on culture in the reduction of hominid biological diversity (Figure 2.4). With the earliest identifiable material culture, over 2 million years ago, there were three genera of hominids in Africa: *Paranthropus*, *Australopithecus*, and *Homo*. Half a million years ago, there was but one hominid species, apparently reliant—and successfully so—on culture for survival: *Homo erectus*. Now, there is a single subspecies of the family of cultural animals: *Homo sapiens sapiens*. Culture, as an effective means of meeting

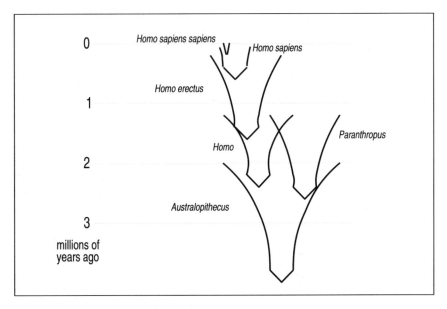

Figure 2.4. Reduction in taxonomic diversity throughout the course of human evolution.

environmental challenges, has had the side effect of making its bearers more biologically homogeneous than other adaptations have made their bearers.

Why has this occurred? Imagine two populations coming into contact, each with different technologies for exploiting the environment, one more successful than the other. They may coexist peacefully, and through intermarriage the knowledge of each population may diffuse to the other. The populations would become more genetically homogeneous, and the more efficient technology would presumably be adopted preferentially by the people—who generally seek to improve their quality of life. Or they have an antagonistic relationship, in which case the technologically superior group would be expected to prevail at the expense of the other population. Either way, the expected long-term pattern would be a decrease in biological diversity along with increasing technological sophistication. Culture is a process of selection, either among the technologies themselves (cultural selection) or among the populations possessing the technologies.

> Anybody could be the first man: it is as easy as to be the first cabbage. To be the first murderer one must be a man of spirit.
>
> —Cain, in George Bernard Shaw's "Back to Methuselah"

Another regularity of particular interest to students of human biology and behavior is probably a consequence of this reduction in biological diversity that has accompanied the evolution of culture. Patterns of biological diversity, specifically the identification of organisms as part of one species and not another, are reflected in elaborate specific-mate recognition systems (SMRS). Each species has distinct means of recognizing its own—based on olfactory, visual, morphological, or other kinds of cues.[25] With the decline of biological diversity among the cultural primates, however, another kind of diversity has replaced the SMRS: the fragmentation of people into cultures, each of which has its own language, traditions, lifeways, and appearances.

Obviously all humans are reproductively compatible with all other humans. But cultures seem to carry on a function in human systems analogous to the role played by the SMRS in general biological systems: they mark a person as being a member of one group, and not another group. This sense of group identification—and group contrast—is a universal property of human societies. Along with the reduction in *biological* distinctiveness that characterizes human evolution apparently comes an increase in *cultural* distinctiveness.

One of the major regularities in the analysis of human diversity is that these group differences are of such great importance that they are often widely mis-perceived to be biological in origin. The direct implication of assigning biological causes to cultural variation is that it serves to demarcate the groups in question in a much deeper way than they would ordinarily be differentiated. Indeed, the difference between two human groups seems tantamount to the difference between two species of animals. The confusion of biological with cultural diversity is the most broad and persistent problem in the study of humans. It serves, however, to highlight the *importance* of group divisions and identifications to members of our species.

NOTES

1. Medawar ([1963] 1991).
2. Landau (1991:10–11).
3. Krimbas (1984).
4. "And God said: 'Let the waters swarm with swarms of living creatures, and let fowl fly above the earth in the open firmament of heaven.' And God created the great sea-monsters, and every living creature that creepeth, wherewith the waters swarmed, after its kind; and God saw that it was good. . . . And God said: 'Let the earth bring forth the living creature after its kind, cattle, and creeping thing, and beast of the earth after its kind.' And it was so. And God made the beast of the earth after its kind, and the cattle after their kind, and every thing that creepeth upon the ground after its kind; and God saw that it was

good. And God said: 'Let us make man in our image.' " (Genesis 1:20–26).

5. Paley (1802).
6. Huxley (1932).
7. Bateson ([1905] 1928), Hooton (1930b), Gould and Lewontin (1979).
8. In fruitflies, interestingly, a male indeed passes on the entire chromosome he inherited, for there is no crossing-over in male fruitflies. Though many of the processes of heredity can be extrapolated from flies to humans, it is important to note that there can be profound differences.
9. Provine (1971).
10. Stern (1943), Crow (1988).
11. Chetverikov ([1926] 1961), Fisher (1930), Wright (1931), Haldane (1932).
12. Dawkins (1986).
13. Bowler (1983).
14. Leakey (1959), Leakey, Tobias, and Napier (1964), Susman (1988).
15. Andrews (1992), Dean and Delson (1992).
16. Johanson and White (1979), Leakey and Hay (1979), Susman et al. (1984), Boaz (1988).
17. Dart (1925), Gregory and Hellman (1938).
18. Broom (1938), Grine (1988).
19. Hill and Ward (1988), Wood (1991, 1992), Tobias (1992), Bilsborough (1992).
20. Hill, Ward, Deino, Curtis, and Drake (1992).
21. Johanson et al. (1987), Hartwig-Scherer and Martin (1991).
22. Weidenreich (1947), Coon (1962) Thorne and Wolpoff (1981), Wolpoff, Wu Xinzhi, and Thorne (1984).
23. Howell (1957), Stringer and Andrews (1988), Brauer and Mbua (1992).
24. The exception to this generalization occurs in an industrialized society, where in the interest of going "back to nature" some people have chosen to behave more like nonhuman primates and consume the placenta, in the paradoxical belief that this is a "natural" human behavior (Travathan 1987:106).
25. Godfrey and Marks (1991).

3

Physical Anthropology as the Study
of Human Variation

A duality of thought in the study of physical anthropology can be traced to the 18th century in the works of the Count de Buffon and Carolus Linnaeus. Both were interested in the works of Nature; but the former was interested in describing and explaining it; while the latter was interested principally in classifying it. This approach carried over into their studies of human variation: Buffon described it and invented a theory of microevolution to explain it; while Linnaeus classified humans into subspecies. This duality of thought remains into the 20th century, when, largely due to the reaction against Carleton Coon's 1962 book, physical anthropology moved from a "Linnaean" to a "Buffonian" approach to human variation.

Two fundamentally different ways of studying human variation emerged as part of the general conflict between the views of the 18th-century naturalists, Buffon and Linnaeus. In the 1735 first edition of his work *System of Nature*, Linnaeus placed humans, monkeys, and sloths within the "Anthropomorpha", yet had said nothing about subdivisions of *Homo*. In the 1740 second edition, he added the four geographical subdivisions that would remain through the last (12th) edition: white Europeans, red Americans, yellow Asians, and black Africans.[1]

According to Linnaeus, the varieties of the human species were simply categories below the species level, similar in kind if not in magnitude to the zoological genera in an order, or the zoological orders in a class. Thus his goal was simply to establish what the natural categories of the human species were. Modern systematics takes the tenth edition of *System of Nature* (1758) as its starting point, and therefore it is worthwhile to see how Linnaeus subdivided humans in that most influential edition of his work. After briefly noting that humans were exclusively diurnal and widespread, Linnaeus reports that "in the wild" they were four-footed, hairy, and unable to speak—based on the reports of aban-

doned "wild children" found in
the forests. He then recognized
five natural categories of humans.
Linnaeus listed the four geographical subspecies of humans
(Americanus, Europaeus, Asiaticus, and Afer), accompanied by
terse descriptions of their appearances and personalities.
Thus, *Homo sapiens americanus*
was "red, ill-tempered, subjugated. Hair black, straight, thick;
Nostrils wide; Face harsh, Beard
scanty. Obstinate, contented, free.
Paints himself with red lines.
Ruled by custom." *Homo sapiens*

Figure 3.1. Carolus Linnaeus.

europaeus was "white, serious,
strong. Hair blond, flowing. Eyes blue. Active, very smart, inventive.
Covered by tight clothing. Ruled by laws." *Homo sapiens asiaticus* was "
yellow, melancholy, greedy. Hair black. Eyes dark. Severe, haughty,
desirous. Covered by loose garments. Ruled by opinion." And last (and
obviously least) *Homo sapiens afer*: "black, impassive, lazy. Hair kinked.
Skin silky. Nose flat. Lips thick. Women with genital flap; breasts large.[2]
Crafty, slow, foolish. Anoints himself with grease. Ruled by caprice."[3]

Linnaeus's fifth subspecies was a grab bag called *Homo sapiens mon-
strosus*, constructed to accommodate the large Patagonians and small
Alpines, the cone-heads of China and Flatheads of Canada,[4] as well as
other remote, deformed, or imaginary people. Curiously, though the
other four subspecies are principally geographically defined, the fifth is
not. The fact that the last subspecies was a geographical hodgepodge
should impress upon us that while Linnaeus's classification of humans
superficially *looks* like the way one might classify mice or clams into
geographical races, it is not. It is rather based on socio-cultural criteria
that correlated only loosely with those geographical criteria: the descriptions Linnaeus gives of each subspecies reinforce this conclusion. He
was not classifying humans as one would classify mice: rather, he was
using broad generalizations and value judgments about personality,
dress, and custom, to classify the human species.[5]

Linnaeus was certainly not the first to divide humans up into large
distinct groups. But he was the first to make it scientific. The Count de
Buffon, rival of Linnaeus, also used behavioral and cultural generalizations in his discussions of human diversity. The difference, however, is
that Buffon was explicit in his rejection of *classification* as the goal of the

study of diversity. Thus, rather than establish precisely how many groups of humans there were, and what they were, Buffon sought to describe and explain the diversity encountered within the human species. In his *Varieties of the Human Species* (1749), Buffon did not divide the species into four, or any set number, of subspecies. Rather, he presented the reader with a travelogue, describing the natives of many remote places, both physically and culturally: Eskimos, Lapps, Tartars, Chinese, Japanese, Siamese, Javanese, Filipinos, New Guineans, Indians, Persians, Arabians, Egyptians, Turks, Swedes, Russians, Ethiopians, Senegalese, Congolese, Hottentots, Madagascarans, North Americans, Caribbeans, and South Americans.

Buffon was indeed trying, however primitively, to derive some sort of historical relationships among these peoples by virtue of their resemblances to one another. Yet Buffon was not asking the questions, How many races are there? and What are they? as Linnaeus and his intellectual descendants did and would. Rather, he tried to ask the questions, How is variation in the human species patterned? and How did it come to be this way?

Buffon's approach to the study of human variation was not, however, the one that prevailed. Though his work was very widely read and influential among the educated public, Buffon's reputation among scientists was quite thoroughly eclipsed by that of Linnaeus. The result was that along with the recognition of a nested hierarchy in nature came an emphasis on classifying all creatures, at all taxonomic levels. And as Linnaeus had done for the human species, putting genera into orders, species into genera, and sub-species into species, so too did his successors.

The growth of systematics obviously was a great advance for zoology in general. The work of Linnaeus was a milestone above all in the triumph of naturalism—the scientific philosophy that placed humans within the sphere of other life, and sought to explain humans as simply a special case of terrestrial life. But noting that Linnaeus was unable to classify subdivisions of the human species in the same professionally biological manner as he classified those of other species, it may be asked in retrospect whether this was indeed an advance for *anthropology*.

The answer provided by the subsequent growth and maturation of

Figure 3.2. The Count de Buffon.

anthropology is now relatively
clear: "How many human subspecies are there, and what are
they?" has led anthropologists

| To define is to exclude and negate. |
| —Jose Ortega y Gasset |

down one of the blindest alleys in the history of modern science. The
question ignores the cultural aspect of how the human species is carved
up; it ignores the geographically gradual nature of biological diversity
within the human species; and it has a strong anti-historical component
in its assumption that there was once a time when huge numbers of people, distributed over broad masses of land, were biologically fairly
homogeneous within their group and different from the (relatively few)
other groups

The fundamental question in physical anthropology from 1758 to
about 1963 (spanning Linnaeus' tenth edition of *System of Nature* to Carleton Coon's *The Origin of Races*) was anti-anthropological, anti-biologi-
cal, and anti-historical. In searching for the divisions of the human
species as a cardinally biological question, the question assumed and in
turn legitimized the proposition that the human species could actually
be divided into a small number of basic biological groups. This assumption is ultimately what students of human diversity owe to Linnaeus.

This is not at all to suggest that Buffon articulated no disparaging
value judgments about other peoples. Buffon was, after all, a French
nobleman and a *philosophe*, and imagined himself with his peers atop the
Great Chain of Being. He could write that the "[K]almucks, who live in
the neighborhood of the Caspian Sea ... are robust men, but the most
ugly and deformed beings under heaven." Or that "[on either side of the
Senegal River,] [t]he Moors are small, meagre, and have a pusillanimous
aspect; but they are sly and ingenious. The Negroes, on the contrary, are
large, plump, and well-made; but they are simple and stupid."[6] Clearly
Buffon was no less an uncritical transmitter of the social prejudices of
his time and culture than was Linnaeus. But Buffon, in acknowledging
the divisions between peoples, could still glimpse enough of the subjugation wrought by Europeans to introduce a passionate digression in his
discussion of the peoples of Africa:

> They are therefore endowed, as can be seen, with excellent heart, and pos-
> sess the seeds of every virtue. I cannot write their history without address-
> ing their state. Is it not wretched enough to be reduced to servitude and
> to be obliged to labor perpetually, without being allowed to acquire any-
> thing? Is it necessary to degrade them, beat them, and to treat them as ani-
> mals? Humanity revolts against these odious treatments which have been
> put into practice because of greed, and which would have been reinforced
> virtually every day, had our laws curbed the brutality of masters, and
> fixed limits to the sufferings of their slaves. They are forced to labor, and

yet commonly are not even adequately nourished. It is said that they tolerate hunger easily, that they can live for three days on a portion of a European meal; that however little they eat or sleep, they are always equally tough, equally strong, and equally fit for labor. How can men in whom there rests any feeling of humanity adopt such views? How do they presume to legitimize by such reasoning those excesses which originate solely from their thirst for gold? But let us abandon those callous men, and return to our subject.[7]

Further, in failing to stress the subspecific divisions of the species, Buffon succeeds in emphasizing the unity of the species. For example, in his work on *The Degeneration of Animals* (1766), he develops a theory of microevolution largely to account for the fact that from a common stock, comprising a single species, humans exhibit such an obvious array of diverse forms that some historical processes of biological divergence must be in operation.[8]

Buffon finds three possible causes for change in any species: climatic temperature, nature of the food, and the evils of slavery. Climatic temperature, he concedes, requires a long time to take effect; even longer for the food, the source of "organic molecules" which are organized by the "internal mold" of a species as it grows into its prescribed form. According to Buffon, the most important microevolutionary effect of slavery is its removal of the organism from its native habitat, the climate and food to which it is accustomed; he uses the term *slavery* to refer as well to the domestication of animals. (Thus Buffon makes another subtle but radical break with more traditional zoology here: he uses humans as an example or illustration of a zoological principle alongside others from the animal kingdom—his own triumph for naturalism.) Buffon goes on to suggest that the climatic effects upon racial characteristics could be tested by transporting some Senegalese to Denmark and seeing how long (maintaining strict endogamy) it would take the Danish climate to turn them white.

In sum, both Buffon and Linnaeus mixed cultural and biological data in their descriptions of the human species, and both rendered inappropriate and naive value judgments in their descriptions. But Buffon's approach to the problem of human variation was inquisitive, descriptive, analytical, experimental; that of Linnaeus was classificatory. Both approached the human species as a part of nature. It was, however, the biology of Linnaeus that held sway, and his approach that came down as *the* scientific approach to human variation.

Johann Friedrich Blumenbach inherited the mantle of human taxonomy. Though following Linnaeus in principle, Blumenbach used strictly anatomical characteristics to define his races of the human species. In the first (1775) edition of his *On the Natural Variations in Humankind*, Blumenbach followed Linnaeus in specifying four "varieties of mankind." He was careful to note that "one variety of mankind does so sensibly pass into the other, that you cannot mark out the limits between them"—and then proceeded to mark out the limits between them: first, Europe, West Asia, and part of North America; second, east Asia and Australia; third, Africa; and fourth, the rest of the New World. In the second edition (1781), Blumenbach redefined the categories to form five varieties, subsuming: (1) Europe, West Asia, North Africa, Eskimos of the New World; (2) East Asia; (3) sub-Saharan Africa; (4) non-Eskimos of the New World; (5) Oceania. Finally, in the third edition of his work (1795), Blumenbach gave names to these varieties: (1) Caucasian, (2) Mongolian, (3) Ethiopian, (4) American, and (5) Malay. Still maintaining that human variation is principally of a *continuous*, not *discrete* nature, and therefore that classifications of the species are fundamentally arbitrary, Blumenbach specified the five named varieties as the principal divisions of the human species. This time, he classified Eskimos as "Mongolian" rather than as "Caucasian."

Blumenbach's descriptions were now strictly physical, not behavioral or cultural (as those of Linnaeus had been), but they were still stereotypes:

Caucasian variety: Colour white, cheeks rosy; hair brown or chestnut-coloured; head subglobular; face oval, straight, its parts moderately defined, forehead smooth, nose narrow, slightly hooked, mouth small. The primary teeth placed perpendicularly to each jaw; the lips (especially the lower one) moderately open, the chin full and rounded.

Mongolian variety: Colour yellow; hair black, stiff, straight and scanty; head almost square; face broad, at the same time flat and depressed, the parts therefore less distinct, as it were running into one another; glabella flat, very broad; nose small, apish; cheeks usually globular, prominent outwardly; the opening of the eyelids narrow, linear; chin slightly prominent.

Ethiopian variety: Colour black; hair black and curly; head narrow, compressed at the sides; forehead knotty, uneven; malar bones protruding outwards; eyes very prominent; nose thick, mixed up as it were with the wide jaws; alveolar ridge narrow, elongated in front; the upper primaries obliquely prominent; the lips (especially the upper) very puffy; chin retreating. Many are bandy-legged.

American variety: Copper-coloured; hair black, stiff, straight and scanty; forehead short; eyes set very deep; nose somewhat apish, but prominent; the face invariably broad, with cheeks prominent, but not flat or depressed; its parts, if seen in profile, very distinct, and as it were deeply chiselled; the shape of the forehead and head in many artificially distorted.

Malay variety: Tawny-colored; hair black, soft, curly, thick, and plentiful;

head moderately narrowed; forehead slightly swelling; nose full, rather wide, as it were diffuse, end thick; mouth large, upper jaw somewhat prominent with the parts of the face when seen in profile, sufficiently prominent and distinct from each other.[9]

What was Blumenbach describing? He certainly recognized that humans varied in a manner that defied discrete categorization. Nevertheless, he felt that by describing a small number of typical human strains, the entirety of the species could be accommodated as simply variations on each of these themes. This Platonic approach to biological diversity (in which natural variation is ignored in pursuit of a transcendent form) was the only approach available in the 18th century, and would not be superseded until a century later, by Darwin.[10] Blumenbach's contribution was to rely on biological (morphological) criteria, divorced from cultural criteria, and from broad inferences about personality. Even so, he erected a classification that he knew did not adequately represent the diversity in the human species. Yet so pervasive was the Linnaean approach to biological diversity that the overarching goal of any student of human diversity would have to be: to establish how many basic categories of humans there are, and what they are.

The history of any scientific endeavor, according to the theories of Thomas Kuhn (1962), involves long periods of "normal science," in which problems are pursued and data are collected, punctuated by conceptual revolutions marked by the generation of new "paradigms."[11] The study of human diversity for nearly two centuries following Linnaeus and Blumenbach involves a period of normal science, in which technical advances, such as the development of statistics, craniometry, and genetics, would add to the store of information on how humans vary biologically. The paradigm it labored under, however, was that of Linnaeus—that somehow the goal of all this data collection involved the determination of a small number of fundamental categories into which all human variation could be collapsed.

But just what the categories represented was not terribly clear. To polygenists, who believed in the fundamental separation of the races, those categories reflected the original differences between divinely created primordial humans (of which Adam and Eve were the progenitors of but one). To some post-Darwinian students of human variation, those categories reflected early divergences of human groups: they represented real clusters of organisms with historical information. To others, the categories reflected merely abstract forms, variants of which could be found in any populations to a greater or lesser extent: they represented arbitrary clusters of traits found at their extremes in certain people or groups of people. And to still others, this paradigm was simply a device or convenience, permitting generalizations about large groups of people.[12] Chapter 4 explains, however, that classifying humans is fundamentally different from classifying snails or flies. First, since humans are both subjects and objects, classification of humans is inevitably a social issue as well as a biological issue, and therefore the recognized categories have power by which to validate inequalities and injustices—which are irrelevant to flies and snails. Second, because of inequalities and injustices, the classification of individual humans takes on significance to those people being classified—which is again not a consequence to the classifier of snails or flies.

In South Africa in 1988, for example, 1,142 people were permitted to change their race by the Race Classification Board at the Ministry of Home Affairs. Yet about 20 times as many people went from "Black" to "Colored" as went from "Colored" to "Black." Obviously the status difference attached to the classification was the source of a degree of self-reflection on the part of many of those who were the objects of the classification.[13]

The Linnaean strain in physical anthropology, the one that saw the goal of the study of human diversity to be the numbering and naming of human groups, came to its crisis in 1963, following the publication of *The Origin of Races* by Carleton S. Coon. The paradigm shift that followed on the heels of this controversy was in many ways Buffonian, documenting and explaining *patterns* of variation and differences among populations. More important, this approach acknowledges the social import of the scientific endeavor as it involves humans. The new paradigm is not a retreat from naturalism; rather it is the fulfillment of naturalism. It acknowledges that what humans do to one another (particularly as scientists) has very different qualities from what humans do to other species. In other words, it recognizes the cultural nature of human science. The science of humans is simply political and value-laden in ways that the science of, say, fruitflies is not: for who would use science to degrade or oppress fruitflies? Thus, the new paradigm seeks to identify, indeed to eclipse, the ideologies that inhere in human science. Humans are still a part of nature, but they are a *different* part of nature, a *cultural* part.

Carleton Stevens Coon earned his Ph.D. at Harvard in 1928, where he became the protegé of Earnest Hooton, the leading physical anthropologist in America. An acknowledged authority on human variation, Coon had written *The Races of Europe* in 1939. He had traveled extensively, taught at Harvard and Penn, and was well known as an intellectual from his appearances on an early television show, "What In The World?" In 1961 he was elected president of the American Association of Physical Anthropologists but resigned after members of the association voted to censure a work called *Race and Reason*, written by a prominent businessman named Carleton Putnam. The book was an anti-integrationist tract that explicitly placed the blame for this subversive idea, and the equally subversive idea of egalitarianism, upon a conspiracy of communists and Jewish anthropologists. The book was even gilded with a foreword by some eminent scientists.[14] Coon, seeing little worthy of objection in Putnam's book,[15] published his own summary of a life's work in biological anthropology the following year: his monumental *The Origin of Races*.[16] It is

Figure 3.3. Carleton Coon.

this work that triggered the paradigm shift in anthropology, the overthrow of the Linnaean approach in the study of human diversity.

Coon divided the world's people into five races, splitting the Africans into two and lumping the Native Americans in with the Asians: Caucasoid, Mongoloid, Australoid, Negroid, and Capoid. Coon's radical assertion was that these five (not four, not six) races/subspecies of modern humans were identifiable in the Middle Pleistocene, as the same five races in *Homo erectus*. Further, they each evolved into *Homo sapiens*, whites first. Though Coon's book lacked the overt social program of Putnam's (for its subject was physical anthropology, not politics, as Putnam's explicitly was), it certainly contained a scientific validation for political action. That justification was part and parcel of the Linnaean approach to human variation, which had been an unquestioned assumption for two centuries: that humans could be clustered into a small number of discrete categories that in turn reflect aspects of the fundamental biology of the people that compose them. In essence, as the biology of different species makes a member of one species different from another, so the biology of different subspecies of humans makes each different from the others. Thus, the continuous quality of human variation that had even impressed Blumenbach as giving the lie to the classification of humans into discrete races was now wholly engulfed by the drive to cluster humans into a few large groups with different basic natures.

The Origin of Races was greeted with considerable skepticism by an anthropological community that was now appreciating the social significance of what its members had thought was "objective" biological work, and was beginning to question its fundamental assumptions. In this instance, a book that was at best anachronistic catalyzed a shift in the perspective of students of human diversity. Critics within the anthropological community doubted the ability to identify five discrete subspecies of living humans, the ability to identify five discrete subspecies of *Homo erectus*, and the possibility of linking them up in ancestral-descendant relationships.[17] The immediate result was a dramatically explicit change in focus, a change that had already been proceeding for the post–World War II generation of biological anthropologists. Programmatically, the criticism engendered by *The Origin of Races* turned anthropologists away from the questions addressed by the book, and toward questions of human variation as local adaptation, toward describing patterns of variation across the species, and toward the study of diversity within natural populations.

The time was also right for such a shift. Though for centuries the Linnaean paradigm had been unquestioned in biological anthropology, largely because it had proven so successful in general zoology and botany, its application to humans was now seen to be significantly limited. First, there was little agreement on how many groups to recognize, what they were, and how to diagnose them—a sign that the categories under analysis might not be natural.[18]

Second, the monarchies of the 18th century that had been dominated by concepts of hereditary rank, and the accordance of rights on the basis of hereditary class, had in the 20th century been superseded by democracies, which accorded rights explicitly to citizens as individuals. Yet even in the model democracy of the United States, the classification of citizens often served in practice to deny rights to citizens as group members that in principle were guaranteed to them as individuals. Consequently, the civil rights movement of the early 1960s emerged to ensure the rights of individuals, regardless of group membership or identification. A science of race, the traditional biological basis for according people unequal rights, was therefore outmoded. Once equal rights were established and enforced, the science of race would be as irrelevant as alchemy or the geocentric solar system. Social and political realities had made the science of race moot: the very purpose of trying to determine a single-digit number of human subspecies no longer had a point.

This was the social matrix into which Carleton Coon's work was introduced, and roundly rejected. Not only was it empirically unsound, but it was pointless as well. The alternative approach acknowledged the apportionment of humans into populations, as had Buffon in the 18th century, but denied biological reality to the higher-order clusterings of populations, which had been an integral part of the Linnaean approach.

A change in focus for biological anthropology, away from the Linnaean paradigm, was evident in the Cold Spring Harbor Symposium of 1950: "Origin and Evolution of Man." William Howells, for example, noted the absence from the study of human evolution of a fit between the study of evolutionary processes and products along the lines that had been achieved in zoology. Likewise, Joseph Birdsell noted that although the study of variation among groups of humans was "not bankrupt, it will require assistance in bridging the awkward gap between its descriptive phase of development and the new analytical phase lying ahead."[19]

Taking a cue from the biological synthesis of that generation, one culmination of which was the "New Systematics,"[20] and which was widely cited at the Cold Spring Harbor Symposium, Sherwood Washburn announced the "New Physical Anthropology" in 1951. Washburn explicitly emphasized a perspective wherein "process and the mechanism of evolutionary change" would replace the archaic paradigm, the one concerned with "sorting the results of evolution," and which had been "static, with emphasis on classification based on types."[21] In 1962, when Carleton Coon's book was published, Washburn was president of the American Anthropological Association, and used the presidential address to contrast the "new physical anthropology" with the old, specifically in the area of race.

> The concept of race is fundamentally changed if we actually look for selection, migration, and study people as they are (who they are, where they are, how many there are). . . .
> Since races are open systems which are intergrading, the number of races will depend on the purpose of classification. . . .
> It is entirely worth while to have a small number of specialists, such as myself, who are concerned with the origin of gonial angles, the form of the nose, the origin of dental patterns, changes in blood-group frequencies, and so on. But this is a very minor, specialized kind of knowledge.[22]

In emphasizing both the historical study of the differences among human groups and their relation to the processes that generated them, and de-emphasizing the classificatory aspects of the study of human variation, Washburn was pronouncing the end of a scientific tradition that reached back to the 18th century. The Linnaean paradigm in physical anthropology had been the culmination of a philosophy of naturalism that regarded humans as just another species, and thereby deserving of treatment by scientists *as* just another species. The fact that zoological treatment resulted in social complications when applied to humans, such as the abrogation of rights on the basis of group inclusion, was now taking on a significance it had not previously been accorded. The excesses that needed to be rectified, clearly, were the result of the *failure* to consider humans as the only political, economic, and moral species. Unlike other zoological classifications, a classification of humans was important not only to the subject (the classifier) but to the objects as well. And particularly if the categories were nebulous and arbitrary in the first place, the usefulness of the entire approach could now be called into question.

The result involved the entrenching of a new paradigm, the Buffonian, which had been resting quietly for 200 years. Buffon's theories were also fundamentally naturalistic, in that humans are regarded as a zoological species. However, the achievement of a classification would not now be the goal of the study of humans. Rather, populations would be the focus.[23] And the new goal would be the analysis of how human groups adapt, how they vary, and what the impact of their histories has been upon their biology.

NOTES

1. Bendyshe (1865).
2. Or "breasts yield copious quantities of milk." Linnaeus was generalizing about the genital flap from reports about Hottentot (Khoi) women. His one sentence ("Feminis sinus pudoris; Mammae lactantes prolixae") has been extensively discussed, particularly in the context of the "Hottentot Venus," who was displayed in Europe half a century later, and ultimately dissected by Cuvier. See Gould (1982), and especially Schiebinger (1993) and references therein.
3. Linnaeus (1758:22).
4. "Macrocephali capite conico: Chinenses. . . . Plagiocephali capite antice compresso: Canadenses" (Linnaeus 1758:22).
5. Linnaeus also divided the existing descriptions of apes into the more anthropomorphic and less anthropomorphic. The former formed the basis for an imaginary second species of humans, *Homo troglodytes*; the latter formed the basis for a species of nonhuman primates, *Simia satyrus*.
6. Buffon, "Variétés dans l'espèce humain" (1749), in Buffon (1749–1804); Buffon ([1749] 1812:382).
7. Buffon, "Variétés dans l'espèce humain" (1749), in Buffon (1749–1804); Buffon ([1749] 1812:394).
8. Buffon, "De la dègènèration des animaux" (1766), in Buffon (1749–1804).
9. Blumenbach ([1795] 1865:265–66).
10. Mayr (1959).
11. Kuhn (1962).
12. Vallois (1953); Stanton (1960).
13. Carlin (1989).
14. Putnam (1961).
15. Coon (1981:334).
16. Coon (1962).
17. Montagu (1963), Birdsell (1963), Roberts (1963). This feeling was not unanimous, it should be noted; the book received several polite and positive reviews. For a revisionist spin, see Shipman (1994).

18. Montagu (1941).
19. Howells (1950), Birdsell (1950).
20. Huxley (1940).
21. Washburn (1951:298).
22. Washburn (1963:527).
23. Thieme (1952).

4

The History of Biology and the Biology of History

Attempts to grapple with the differences among human groups have tended to be intimately linked with both the notion of the Great Chain of Being (which promotes notions of racial superiority), and the economic suppression of the lower classes (which is rationalized by notions of racial superiority). Nineteenth-century theories of history failed to divorce racial from cultural history, and were undermined by the development of the culture concept, employed by Franz Boas and his students.

By the end of the 18th century most reflective Europeans appreciated that things had not always been as they are. Social and economic changes—apparently advancements—had brought a downfall to the hereditary monarchies that had seemed stable, indeed divinely ordained, only a few generations earlier. That century saw as well the framing of historical approaches to the origins of contemporary institutions. Biologists of the 18th century were questioning the stability of the earth and its species, and speculating upon the origins of new species. Likewise, the origins of European language, marriage, laws, and religion were being treated for the first time as history rather than as eternal verities.[1]

Thus, many different avenues of change were being considered in parallel. One could freely speculate on our emergence as people from non-people ancestors, as Lamarck did; and one could speculate on our emergence as civilized people from non-civilized ancestors, as many of his contemporaries in social philosophy did. Jean-Jacques Rousseau, for example, reasoned from man in a state of nature, "satisfying his hunger under an oak, quenching his thirst at the first stream, finding his bed under the same tree which provided his meal," to the civilized man of 18th-century Europe in his *Discourse on the Origins of Inequality* (1755).[2]

Rousseau, however, was astute enough to recognize that historical speculations about the human species involved answering two distinct kinds of questions, about the origins of human biology and about the origins of contemporary society.

I shall not pause to investigate in the animal what man must have been at the beginning in order to become in the end what he is. I shall not ask whether man's elongated nails were not originally, as Aristotle thought, hooked claws, whether his body was not covered with hair like a bear, or whether walking on all fours, with his eyes directed towards the earth and his vision confined to several paces, did not shape the character and limits of his ideas. . . . Thus, . . . without regard to the changes that must have taken place in man's configuration, both inwardly and outwardly, as he put his limbs to new uses and nourished himself on new kinds of food, I shall suppose him to have been at all times as I see him today, walking on two feet, using his hands as we use ours, casting his gaze over the whole of nature and measuring with his eyes the vast expanse of the heavens.[3]

Rousseau thus divorces the biological origins of his subject from the cultural origins. His subsequent reasoning about the origins of values and morals is carried on without recourse to the biological origins of the humans possessing those values and morals. In retrospect, we could assist Rousseau in arguing that (1) the values and morals that he was documenting were formed long after the human species had taken its shape; and (2) the processes of biological evolution and cultural evolution are different in mechanism, in time-frame, and in transmission mode to the next generation, so that in spite of sharing a common label ("evolution"), they in fact are categorically distinct phenomena.

Others were not so perceptive as Rousseau. As noted in Chapter 3, Linnaeus used traits like clothing to distinguish his subspecies of humans, as if their modes of dress were part of their constitution. When Linnaean systematics became evolutionary a century later, the implication would be that patterns of culture, such as clothing fashions, were evolving alongside the biological features of those peoples—and might reasonably be considered constitutional or ingrained.

HISTORY AS INBORN PROPENSITIES: ARTHUR DE GOBINEAU

The search for theories of change was a preoccupation of 19th-century scientists. The most notable success, of course, was achieved in biology, by Charles Darwin. But other kinds of changes required an explanation as well—possibly the same explanation, or possibly other explanations. Ultimately, however, the question of how civilization came to develop from primitive barbarism deserved an answer no less scientific than how humans came to develop from primates.

In 1855, the first volume of an *Essay on the Inequality of the Human Races* was published by a French nobleman, Count Arthur de Gobineau. In it, Gobineau sought to explain the rise and fall of civilizations. The

Essay was an intellectual work, and without recourse to metaphysics Gobineau produced "a striking single-cause explanation in fundamentally secular terms."[4] His theory was that civilization rises and falls in proportion to the purity of "Aryan blood" contained within it.

Gobineau's work was the product of the decaying aristocracy, threatened by the new urbanism and egalitarianism of the industrial age. Indeed, it was a monumental preachment against democracy, a perfervid defense of aristocracy and feudalism, an expansion of the vanity of a proud-spirited poet into a 'scientific' interpretation of all civilizations as the creation of a fictitious race of which he imagined himself . . . to be a member.[5] He fancied that the times in which he lived were decadent, and heralded the decline of civilization. Consequently, Gobineau sought to identify whatever general principles might govern such decline with the social forces he observed with contempt around him.

> More than any time in history, mankind faces a crossroads. One path leads to despair and utter hopelessness. The other, to total extinction. Let us pray we have the wisdom to choose correctly.
>
> —Woody Allen

According to Gobineau, there were ten identifiable civilizations in the history of the world, all of which rose as the Aryans brought their inborn tendency to cultivate arts and knowledge to local peoples. The concentration of this gift in the Aryans, coupled with the other, lesser gifts of the indigenous inhabitants, resulted in civilization. As each civilization flourished its members saw themselves as equals, rather than as fundamentally different races possessing fundamentally different gifts, as Gobineau saw them. One reason for the egalitarianism is that they had been interbreeding, and so no longer could recognize the basic inequalities in themselves that Gobineau saw in their ancestors. Thus, as the groups became interbred over time, the genius for civilization became dissipated, as the cosmopolitanism and democracy Gobineau observed in his own era took hold. Civilizations ultimately rose and fell as a result of the purity of Aryan ancestry in their founders, and its preservation in their descendants. The descendants paled by comparison to their ancestors, however:

> [T]hough the nation bears the name given by its founders, the name no longer connotes the same race; in fact the man of a decadent time, the *degenerate* man properly so called, is a different being, from the racial point of view, from the heroes of the great ages. . . . He is only a distant kinsman of those he still calls his ancestors. He, and his civilization with him, will certainly die on the day when the primordial race-unit is so broken up and swamped by the influx of foreign elements, that its effective qualities have no longer a sufficient freedom of action.[6]

Of course, Gobineau was consequently obliged to find Aryans everywhere, from China to Peru. And he did.

Gobineau was certainly not the first to propose that different human groups have different inborn propensities. But his general theory of the rise and fall of civilization *by recourse* to those different inborn propensities of human groups, his isolation of the single group responsible for *all* civilizations, and his identification of cultural decadence and decline with biological admixture, was an original synthesis and made his theory attractive for its simplicity and apparent scholarship. Indeed Jacques Barzun attributes to Gobineau the first popular synthesis between racial thought in anthropology and in history.[7]

These ideas were not particularly highly regarded at the time, but Gobineau's *Essay* would later develop a wide appeal, being translated into English, for example, during the World War I, some sixty years after its original publication. The renaissance of Gobineau's *Essay* is related to the strong popular racialism that took hold at the end of the century. Houston Stewart Chamberlain's *Foundations of the Nineteenth Century*, published in Germany in 1899, was a theory of European history greatly indebted to Gobineau, as was Madison Grant's *The Passing of the Great Race*, published in the United States in 1916.

Though each work put its own particular slant on the subject, all shared the common property of interpreting the history of the world through genetics. In this way, civilization is seen as an organic property, encoded in the hereditary constitutions of organisms, like blue eyes or curly hair. The causes of history were simply reduced to the unfolding of the natural propensity of the specific people involved. In this way, however, nothing is really explained: civilization began because it *had* to, for it was inscribed in the genetic destiny of its bearers.

HISTORY, BIOLOGY, AND THE THEORY OF PROGRESS

Gobineau, as an aristocrat witnessing the decline of aristocratic rule in Europe, naturally saw history through a pessimistic lens, one in which decay and decline were the signs of the times. Democracies rose, to a large extent, at the expense of the aristocracy. Generally, therefore, most people did not see the social changes as quite so threatening. Rather, they saw progress: economic, social, civil, scientific.

Where progress could be seen, those who saw it were usually its beneficiaries. And if progress could be discerned from the past to the present, why not from the present to the future as well? Progress, in fact, becomes one of the leading philosophical themes of the 19th century.

The theme of progress has been explored in several classic works of the history of ideas.[8] For present purposes, however, this theme is of interest insofar as it had an impact upon theories of human biology. If things were in fact getting better, if there were a transcendent direction to human history, several implications could easily follow. First, the human species could be seen as further progressed, or more perfect, than other species; second, European civilization could be seen as further progressed (more perfect) than modes of life elsewhere in the world; and third, civilized Europeans could be seen as further progressed (more perfect) than other kinds of people. All three of these implications are different variants of a Great Chain of Being (representing rankings of species, lifeways, and organisms, respectively), but they are all of a piece with the general vision of progress.

How does improvement work? Assuming there is a goal towards which life and civilization is groping, how did it become established? Is it external, established by God? Is it internal, as postulated for species by Lamarck? If there is a goal towards which we are striving, then is the future pre-destined, and is there a way to know what it is?

Eighteenth- and 19th-century views of change as progress were promoted widely, and often without the overlay of genetic determination. For example, often it was the gross environment that dictated the form of life or of behaviors. According to Buffon, foods contained particular "organic molecules," which influenced the physical appearance of the local people consuming them.[9]

While goal-directed, or teleological, evolution is not now evident in the biological realm, it has always been more apparent in a particular cultural realm, most clearly in technology. Consequently, as archaeology began to uncover the "development" of Western civilization in a sequence of technical improvements rising out of the Stone Age and through the Bronze and Iron Ages, it seemed reasonable to suppose that comparable improvements were occurring in the form of the human species as well. This, of course, led to a critical question for the colonial empires: What was the relation of European Industrial Age peoples to non-European, not-so-far-advanced-as-Industrial-Age peoples? What was the significance of this clear technical superiority? Was it an outgrowth of organic superiority? Or some other kind of superiority?

With the Darwinian revolution came a new scientific theory for comprehending change: competition. Just as competition had led to the emergence of new and diverse life forms, so too might competition be the cause of the diversity of modes of life among human societies. If competition is viewed as the natural, pre-eminent mode by which nature establishes diversity on earth, and extinction is its inevitable by-product, the implications for social phenomena are stunning. Competition, indeed, could now be seen as the mechanism by which nature produced not merely diversity, but progress, as historian Peter Bowler has shown.[10]

The paradox here is that in Darwin's view, competition in the biological realm leads to diversity, *not* progress; progress, however, was so strongly a part of the European world-view that Darwinian competition was simply grafted onto it. The failure to divorce the processes of cultural and biological evolution from each other, the inference of continuous improvement throughout history, and the mechanism of competition combined to produce a general attitude among Europeans that was highly conducive to the global politics of colonial imperialism.

Competition is natural and progress is good, reasoned the colonialists. Extinction is inevitable and necessary for progress to occur. It follows that the contact of peoples should quite reasonably involve extensive acculturation, and the replacement of older lifeways with better ones. Moreover, our own history can answer the historical question of how we got to be in a position of cultural dominance in the first place. Surely the reason we were able to make the rest of the world see things our way is that our ancestors had obviously out-competed all others. This successful competition produced a heartier civilization, and possibly a heartier human organism as well.

The diversity of thought, out of which arose the modern view of the relationship between biology (or "race") and lifeways (or "culture"), involved the interplay of the notions of progress and competition. The works of four influential social scientists demonstrate how the ingredients were recombined: Georges Vacher de Lapouge, who saw social processes intertwined with biological processes in a strictly dog-eat-dog view of "social selection"; Herbert Spencer, who separated social from biological phenomena, but saw them each as progressing by parallel processes of competition; Karl Marx, who saw social progress as independent of biological processes, and for whom competition occurred not so much between individuals as between social classes; and Franz Boas, who undermined the existence of progress in the cultural sphere, and divorced the biology in "race" from the history in "culture".

SOCIAL SELECTION: BIOLOGICAL PROGRESS
AS SOCIAL PROGRESS

The development of the social sciences in the early 20th century involved understanding the different roles of biological and cultural processes in human history. To Lapouge, the history of peoples was indistinguishable from their phylogeny. Traits such as bravery, tractability, and free thought had been bred into (or out of) human groups by the selective agents operating throughout their history.[11] Behavioral differences among peoples would thus have an innate basis: the result of natural selection operating through history, making breeds of humans as fundamentally different as golden retrievers and Dobermans.[12]

Even fairly naive readers, however, saw that this theory did not provide a useful model for why groups of people differed from one another in their behaviors. Mass selection would have had to occur consistently in very large populations over a long period of time in order to create an effect significant enough to explain behavioral differences among human groups. Were there peoples whose history was so monotonous that the same selective agents operated in several successive generations? Were populations so homogeneous that something like "bravery" could be bred out of them without continually reappearing in the next generation? Were human behaviors and attitudes so simple that changing or eliminating them was as simple as retyping or erasing a word? And finally, were these characteristics so innate and immutable that selection could have a significant effect upon them? No, most likely, was the answer to all these questions: how people's lifeways and attitudes came to differ from one another was probably not due to precisely the same forces that had led to the divergences of their physical form.

SURVIVAL OF THE FITTEST:
PARALLEL PROGRESSIVE PROCESSES

Herbert Spencer similarly cast natural selection as the shaper of human differences, but he visualized twin parallel processes operating on biology and society.[13] Competition was the driving force behind progress, both social and biological. Just as the struggle for existence had brought the human species to the pinnacle of the animal kingdom, so had competitive struggle brought industrial England to the pinnacle of the civilized world.

According to Spencer, biological events did not necessarily mold social forms: Rather, competition was pervasive in all spheres, including the biological and social. In both arenas, competition was the agent of evolutionary change. Competition was Spencer's clarion, rationalizing the exploitation of workers by employers, and his writings were rightly perceived as an apology for laissez-faire capitalism. In this theory, however, social evolution was largely distinguished from biological evolution.

Thus, while Spencer's theories were compatible with the idea of intrinsic differences being at the root of behavioral differences between groups, they did not require it. Further, this theory did not justify the

inherited superiority of the aristocracy, but the superiority of the nou-
veau-riche: it was they who had out-competed their rivals and thereby
were entitled to enjoy the fruits of their labor.

Competition—like colonization by the Empire—was therefore seen as
the mainspring of progress. But could the competition that lent validation to the sweatshops and coal mines possibly be progressive? Competition among people as articulated by Spencer was at best an incomplete
theory—for it was not at all clear how the free competition he believed
had brought societal advancement about could also be invoked to facil-
itate further progress— presumably a better life for all people. Indeed
Spencer's theory implied that all people did *not* deserve a better life. The
ones who had a good life deserved it; the ones who didn't have it would
probably never get it, because they lacked the competitive edge.

COMPETITION OF A DIFFERENT SORT:
PROGRESS IN HISTORY WITHOUT BIOLOGY

A different school of thought maintained that the forces that forged
humans from a primate stock were different in kind from those that generated political and social differences among people. For example,
although Karl Marx embraced Darwinian evolution through competition
between individuals as the basis of *biological* history, he envisioned *social*
history as occurring through the competition between economic
classes.[14] If true, this theory implied that the forces that appear to produce progress in the natural world are fundamentally different from
those which produce progress in the social realm.

Outside Europe, however, ideas of class struggle were difficult to
apply. It was not clear, after all, that "uncivilized" non-Europeans had
the same kind of economic stratification as "civilized" Europeans and
Asians. Nevertheless, if history (the chronicle of human events) occurred
by processes that were independent of phylogeny (the chronicle of biological events), then human progress could be understood without any
recourse to biology.

To Marx, people's actions were broadly dictated by their economic
interests. And to the extent that people in similar economic strata had
strongly convergent interests, people tended to act in accordance with
their class interests. But since economic interests cross-cut racial lines,
people of different biological heritage would tend to behave similarly if
they found themselves in the same economic circumstances. Thus,
whatever biological history may have made them different would be
submerged by the particular context that would make their behavior
similar.

And so it appeared to be: the relations between laborer and entrepreneur were similar, regardless of whether the laborers were Teutons or Slavs. Likewise, the general relations between vassal and lord at an earlier period are what shaped European history, and transcended the biological differences between particular sets of vassals and lords in various parts of Europe. Feudalism collapsed throughout Europe because of common economic interests, not because of biological changes in the populations, and it occurred in spite of the biological diversity across Europe. Thus, whatever progress had occurred throughout history, and would presumably continue to occur, was independent of the biological composition of the particular human populations involved. It was dictated by other processes—the general desire of humans to free themselves from subjugation by others, and to live in relative comfort.

THE DIVORCE OF RACE AND CULTURE: PROGRESS AS AN ILLUSION

Franz Boas emerged as the intellectual leader of American anthropology by 1920, and was ultimately responsible for effecting the divorce between theories of biological and social change. He accomplished this by jettisoning the last vestige—the most tenacious one—of the old biological-social thought: the assumption of progress.

While most theories that linked biological to social change throughout the 19th century centered on the explanation of progress, to orthodox evolutionary biologists the very problem was a false one. It was a direct implication of the Linnaean view of nature that the human species was smack in the middle of the natural world; that our particular adaptation was intelligence, but it could not be ranked as an adaptation objectively superior to the swiftness of a cheetah, the strength of a gorilla, or the vision of an eagle. Biological chauvinism, or *anthropocentrism*, often impeded the full appreciation of this fact, but now and again scientists did acknowledge that biological progress was rather more difficult to prove than simply to assert.[15]

Yet progress in human culture remained unquestioned. Even as Marx and his followers attempted to establish a science of history, they still conceived of history as general improvement over the past and

Figure 4.1. Franz Boas.

continuing into the future. The first two decades of this century, however, brought anthropology out of the armchair (in which an earlier generation had read literature with highly variable degrees of credibility) and into the field, where detailed studies of other societies began to undermine the assumption that things had actually been continuously improving in the history of our culture.

The comprehensive concept of culture that had been devised by E. B. Tylor ("that complex whole which includes knowledge, belief, art, morals, law, custom, and any other capabilities and habits acquired by man as a member of society") served Boas well. For although it was fairly clear that there had been progress in the area of technology (World War I had already shown that to be a mixed blessing), could progress also be seen in myths and languages? Or kinship systems?

No, argued Boas. Cultural evolution was not so much consistent progress as a trade-off—progress in one area, matched by retrogression in another:

> The impoverishment of the masses brought about by our unfortunate distribution of leisure is certainly no cultural advance, and the term "cultural progress" can be used in a restricted sense only. It refers to increase of knowledge and of control of nature.
>
> It is not easy to define progress in any phase of social life other than in knowledge and control of nature.[16]

Witnessing exceedingly rapid cultural change in his own time, as well as the large-scale forcible diffusion of Western culture to non-Western societies, Boas appreciated that the pace of cultural change could not be matched by the pace of biological change. Biological evolution, therefore, could not be a cause of cultural evolution, for biology changed too slowly to account for cultural changes. The two processes were not connected in any obvious way, and therefore human cultural history was interpretable only in terms of cultural forces, not biological ones.[17] (A convergent conclusion— that social facts can be explained only by recourse to other social facts—had been reached by the school of French social scientists led by Emile Durkheim.[18])

> Human history becomes more and more a race between education and catastrophe.
>
> —H. G. Wells

Thus, not only was cultural evolution not the history of progress, it was also not the history of race, if that term were taken to refer to some biological aspect of a population. The biological history of the human species would be the study of race; the social history of the human species would be the study of culture. By the 1920s, human history was a history of culture. It might be explicable as the clash of cultures with one another, or even economic subgroups within a single culture, but the clash of races—biologically different populations—was no longer an explanation for the diversity of human cultural forms.

THE CULTURE CONCEPT NUDGES OUT THE RACE CONCEPT

This was not to say that the origin and maintenance of human bio-logical diversity was not an interesting and significant issue, worthy of scientific analysis. Rather, it simply meant that this scientific issue was categorically different from the one in which the student of human social forms was interested. To the extent that different biologies are not the causes of variation in attitudes and values across human popula-tions, human biology could be regarded as a constant in the analysis of human cultural variation and evolution.

Boas, it should be noted, did not hold that there were no significant biological differences among peoples, or that "races" did not exist—only that they did not suffice to explain the variation in attitudes and behav-ior encountered around the world. They were worthy of study, but not as an explanation of cultural differences; those were the result of differ-ent histories, not different biological backgrounds.

To many casual observers, however, this was a counter-intuitive assumption. After all, human biological variation was strongly *correlated* with human cultural variation. How could the fact that the most bio-logically different peoples were also the most culturally divergent peo-ples *not* be a significant causal association?

The vigilance of Boas and his students in divorcing questions of human history from the gene pool was tireless, and needed to be so in the face of such an apparent correlation. The correlation had been obvi-ous to Gobineau in the mid–19th century:

> So the brain of a Huron Indian contains in an undeveloped form an intel-lect which is absolutely the same as that of the Englishman or Frenchman! Why then, in the course of ages, has he not invented printing or steam power? I should be quite justified in asking our Huron why, if he is equal to our European peoples, his tribe has never produced a Caesar or a Charlemagne among its warriors, and why his bards and sorcerers have, in some inexplicable way, neglected to become Homers and Galens.[19]

Gobineau's pompous questions were simply misdirected. The answers were not to be found in the biological structure of the brain, but in the circumstances of history. Caesar, Charlemagne, Homer, and Galen were not average Europeans, to judge by their accomplishments, and their accomplishments are impressive because we have records of them. The

development of writing (which permitted Europeans to recall the existence of those great men) was not a genetic endowment, but a singularity—or at least a rarity—of history.

One attraction of Gobineau's theory, however, is that if the invention
of writing were attributable to a general genetic endowment of genius,
then Gobineau and his readers could presumably lay claim to sharing it.
If it were, on the other hand, only an historical event, then Gobineau
and his readers could lay claim to no part of it, as they would be simply passive inheritors of somebody else's idea many thousands of years
after the fact. They could partake of the greatness of Caesar, Charlemagne, Homer, and Galen, while not necessarily having done anything
"great" themselves. Thus, a theory of history in which great events or
discoveries are explicable as genetic facts also serves to *democratize* the
great event, by locating it to the genome of the reader, who would otherwise have no claim to it.

The first biologist to acknowledge that cultural dominance was simply the result of historical contingencies, and not an indicator of biological superiority, was the English Marxist Lancelot Hogben.

> At a time when we hear so much of the superiority of the Nordic race, it
> may be well to bear in mind the views of those who were preparing the
> ground for the cultural development of Northern Europe when our own
> forbears were little better than barbarians. A Moorish savant, Said of
> Toledo, describing our ancestors beyond the Pyrenees, observed that they
> "are of cold temperament and never reach maturity; they are of great
> stature and of a white colour. But they lack all sharpness of wit and pen-
> etration of the intellect". This was at a time when a few priests in North-
> ern Europe could read or write and when washing the body was still con-
> sidered a heathen custom, dangerous to the believer, a belief that lingered
> on to the time when Philip II of Spain authorised the destruction of all the
> public baths left by the Moors. The Moorish scholars of Toledo, Cordova,
> and Seville were writing treatises on spherical trigonometry when the
> mathematical syllabus of the Nordic University of Oxford stopped
> abruptly at the fifth proposition of the first book of Euclid.[20]

The association between cultures and crania, and by extension genetic
endowments, was not so easily shaken, however. It was certainly more
threatening to see history as capricious rather than as simply the unfolding of biological destiny. Consequently Harvard's Earnest Hooton could
still write in 1946, "I can see no reason why . . . the pygmy should not
have acquired a culture, except inherent lack of mental capacity—which
in terms of gross anatomy means an inferior brain."[21]

More than a century after Gobineau, but with only a little more
sophistication, Carleton Coon would write:

[I]t is a fair inference that . . . the subspecies which crossed the evolution-ary threshold into the category of *Homo sapiens* the earliest have evolved the most, and that the obvious correlation between the length of time a subspecies has been in the *sapiens* state and the levels of civilization attained by some of its populations may be related phenomena.[22]

Coon seems to express here that cultural evolution is an automatic out-growth of biology: driven by large-brained geniuses, cultural evolution was simply a function of how long those large-brained geniuses had been produced by the group in question. Yet by the time his sentences were written, any causal connection between civilization and biology was widely appreciated as a non sequitur.

Cultural history is not an accu-mulation of good ideas thought up by geniuses. It is certainly not evi-dent that times of rapid cultural change are determined by the pro-

> He was a bold man that first [ate] an oyster.
>
> —Jonathan Swift

portion of geniuses born. Indeed, a good idea is but a small step in cul-tural change; the major question that requires explanation is why peo-ple change what they have been doing to adopt the innovation.[23] The history of science shows us that good ideas often come up in the minds of several different people concurrently, which implicates the milieu, rather than the genotype of the thinker, as the major determinant of the idea. Further, there are many reasons why people adopt bad ideas or reject good ideas; it is certainly not the case that a good idea is auto-matically recognized as such and adopted.[24] To explain the events of human history, one needs a theory of culture, and can largely take genetics for granted as a constant in the equation.[25] There always seems to be a person around with an idea when you need one; whatever the limiting factor in cultural evolution may be, it does not seem to be whether people can come up with ideas.[26] If, therefore, individual men-tal processes do not underlie the major features of cultural evolution, then it is reasonable to ignore biology—i.e., to regard it as a constant—in the analysis of cultural processes.[27]

The contribution of the Boas school, which cannot be overestimated, was that it conceptually divorced biological history from cultural his-tory. In refuting the notion of cultural progress, this explanation also undermined the Great Chain of Being, which had been used in two dif-ferent ways. First, if biology *did* cause specific cultural forms, then higher cultures implied "better" races, the bedrock of what is now rec-ognized as the pseudo-science of racial superiority. Second, even if biol-ogy *did not* cause specific cultural forms, the ranking of cultures would imply a basis for ethnocentrism, and afford a rationalization for the exploitation of peoples and suppression of their lifeways.

Either way, Boas's theory of culture undermined the general ranking of people or their particular ways of life, and provided if not specific explanations for things, at least an idea of the intellectual realm in which explanations would likely be found.

NOTES

1. Harris (1968), Trautmann (1992).
2. Rousseau ([1755] 1984:81).
3. Ibid.
4. Biddiss (1970:113). See also Poliakov (1974), Banton and Harwood (1975).
5. Hankins (1926:4).
6. Gobineau ([1854] 1915:26).
7. Barzun ([1937] 1965).
8. Butterfield ([1931] 1965), Bury (1932), Nisbet (1980), Lasch (1991).
9. Fellows and Milliken (1972), Bowler (1973a), Eddy (1984).
10. Bowler (1983).
11. Vacher de Lapouge (1896, 1899), Barzun ([1937] 1965), Ruffie (1986).
12. This tenacious idea resurfaced not too long ago in a widely publicized comment by television sports figure Jimmy "The Greek" Snyder. Snyder speculated on innate differences between blacks and whites as having resulted from differential breeding regimens during the course of slavery in America. Vacher de Lapouge himself is widely regarded as the father of modern anti-Semitism, for his writings explicitly ascribe ingrained behavioral attributes to peoples, and especially to the Jews.
13. Spencer (1896), Sumner ([1906] 1940), Hofstadter (1944), Jones (1980).
14. Engels ([1880] 1940), Haldane (1940), Padover (1978), Levins and Lewontin (1985).
15. Nitecki (1988).
16. Boas ([1928] 1962:220).
17. Stocking (1968).
18. Harris (1968), Lukes (1973), Hatch (1973).
19. Gobineau ([1854] 1915:37).
20. Hogben (1932:213–14).
21. Hooton (1946:159).
22. Coon (1962:ix–x).
23. Hallpike (1988).
24. Barnett (1953), Basalla (1988).
25. Kroeber (1923), Linton (1936).
26. Spier (1956).
27. White (1949). White wrote extensively against the Boas school for its failure to grapple with the principles and general properties of cultural evolution; nevertheless, his own ideas owe a largely unacknowledged debt to the distinction between race and culture drawn by the Boasians.

CHAPTER

5

The Eugenics Movement

The only major influence scientists have had on social legislation came in the 1920s, when eugenicists successfully campaigned for involuntary sterilization of "unfit" people and for the restriction of immigration. The theory was an attempt to remedy social problems through biological means, and though tempting in its simplicity, it was conceptually flawed and failed to solve society's problems, which require social, not biological solutions.

Probably the most instructive episode in the history of the study of human biology was the eugenics movement, which originated in late-19th-century England, flourished in America between about 1910 and 1930, and died out with World War II. It is out of the eugenics movement that the study of human genetic variation was born.[1]

But tracing the eugenics movement is not simply an exercise in the history of social thought. It is paradigmatic for the scientific study of human biology. We see in the eugenics movement how *any* study of human biology encodes social values, a situation that the study of clam biology or fly biology does not have to face. We see how scientists expounded on subjects they knew little about, derived results we can now see as thoroughly unjustified, and validated their own social prejudices with the "objectivity" of science. While the eugenics movement was certainly an embarrassing episode in the history of biology, one would be wrong to ignore it as an aberration or an exception. It isn't the exception: it encapsulates the "rule." Studying humans can't be done as dispassionately as studying clams, for there is far more at stake. Therefore the levels of criticism and scholarship must be higher, and the stories that emerge must be subjected to more intense scrutiny from the scholarly community.

A SIMPLE PLAN FOR MAKING LIFE BETTER

The work of Francis Galton in the latter part of the 19th century established as a major goal of biology the betterment of the human species. In an age that valued its aristocracy and thrived on its ability to exploit its colonies, Galton's ideas were taken very seriously. His work involved literally the origin of modern statistical analysis, and was directed toward a presumably humanitarian goal: improving the lot of the people on earth.

The eugenics program was formulated as an outgrowth of Darwinism. Given the persuasive analogy Darwin could make between humans breeding animals to establish various characteristics in populations, and nature breeding species with different characteristics, Galton simply reasoned that humans could be selectively bred for favorable traits as well.

Selective breeding is only effective for inborn attributes, however, and Galton's first task was to show that favorable traits were indeed inborn in people, which he did in his 1869 book, *Hereditary Genius*. Facile in the extreme (Galton claimed to demonstrate that "prominence" is inborn, because prominent people, graded by an alphabetic scale, appear to be derived from prominent families), the empiricism in the work is of little value. Of greater importance are the ideas that underlay the empirical findings:

> that a man's natural abilities are derived by inheritance, under exactly the same limitations as are the form and physical features of the whole organic world.[2]
>
> that the men who achieve eminence, and those who are naturally capable, are, to a large extent, identical.[3]
>
> that the average intellectual standard of the negro race is some two grades below our own.[4]
>
> that the average ability of the [ancient] Athenian race is, on the lowest possible estimate, very nearly two grades higher than our own—that is, about as much as our race is above that of the African Negro.[5]

Galton proceeded to rank English (and other European) men of note by the quality of their reputations, F and G being unique and eminent, and A being mediocre. He then provided thumbnail biographies of eminent people who had eminent relatives. Given the connection between reputation and ability, the inheritance of ability, and the linear scale upon which to assess them, Galton could persuasively argue for the desirability of moving as far up the scale as possible: certainly a nation of Ds would be better than a nation of Bs!

Yet lacking in his analysis was a fundamental scientific necessity: a control. Any body of data requires something with which to compare it, in order to assess whether the explanation for the pattern apparent within the data is valid or not. In this case, Galton would have needed to show not only that men of high reputation have relatives of high reputation, but also that they outnumber men of high reputation who do *not* have relatives of high reputation. Galton constructed a considerably arbitrary list of 37 thumbnail sketches of eminent "Literary Men" with prominent relatives, including Friedrich Carl Wilhelm von Schlegel, Seneca, the Marquise de Sévigné (a woman, as it suited his purpose), and Mme. Anne Germaine de Staël (likewise)—but not Suetonius, Spinoza, or Sir Walter Scott; likewise, as "Poets": Mackworth Praed, Jean Racine, and Torquato Tasso—but not William Blake, John Keats, or William Shakespeare.

> May men say, "He is far greater than his father," when he returns from battle.
>
> —Homer, The Iliad

When, several decades later, biologist Raymond Pearl fell away from the eugenics movement Galton had founded, he redid Galton's study and came to opposite conclusions. For example, Shakespeare was surely a G (a one-in-a-million guy), but what about his father?

> As a matter of fact [he] was the greengrocer and butcher of the town, doubtless an amiable and useful citizen, but after all probably not greatly different from greengrocers and butchers in general. Whereas Shakespeare himself was really a quite superior man in his chosen line of endeavor.[6]

Pearl went on to use the *Encyclopaedia Britannica* as a source of information on prominence, and to ask: Given poets of reputation sufficient to be mentioned in the *Encyclopaedia*, how many of their fathers earned a listing of their own? Of the 72 poets listed whose fathers were known, only three had fathers of enough repute to receive a separate listing in the encyclopedia. Pearl concluded that neither the parentage nor the offspring of eminent individuals is particularly noteworthy, and therefore that genetics plays a far smaller role in achieving notoriety than Galton thought.

Galton's work, however, managed to support several commonsense assumptions of the European intellectuals and gentry: They were constitutionally superior to the "common man," they must be derived from good stock, and people and groups could be linearly ranked along a single scale—with themselves, and their race, at the top.

Galton's originality lay in two areas. First, in his use of quantification: as noted, statistics developed largely as a result of this work. Two other major figures in the development of statistics, Karl Pearson and Ronald Fisher, acquired their interest in the manipulation of scientific data through an interest in eugenics.[7] Galton's second original contribution was in grafting contemporary advances in biology (namely, natural selection) on to old social prejudices, and using the newly emerged theory of how species change adaptively to address the old problem of how to change society for the better.

MENDELISM IN EUGENICS

Throughout the 19th century, the possibility of breeding better citizens could only be discussed in the absence of a detailed theory of heredity. Consequently the English eugenics movement was tied to the school of biometry, the statistical analysis of phenotypic inheritance. With the rediscovery of Mendel's laws, however, and the recognition that hereditary information was passed on in discrete units, eugenics in the 20th century (and particularly in its American incarnation[8]), came to be concerned with the inheritance of discrete units of socially desirable traits or, conversely, with the elimination of the alleles for their alternative undesirable states, such as feeble-mindedness and licentiousness.

Here again the core assumptions and programs *preceded* the discoveries about biology (much as Galton's assumptions preceded the Darwinism on which they were presumably founded). The program hardly changed, though its biological underpinnings had been rewritten; but now the eugenics movement could claim as its basis the new science of Mendelian genetics.

Anthropologist Leslie White analyzed the scope of science in an essay in 1947, and observed that the earliest-maturing sciences were the ones whose subjects were sufficiently far removed from the observer that they could be studied with the greatest dispassion—such as astronomy and chemistry. The latest sciences to mature, the "soft" sciences, are the ones that concern themselves most explicitly with the questions of who we are and why we do things, and the ones that are most difficult to study with sufficient rigor and dispassion.[9]

Connecting the hard and soft sciences, however, is a bridge of pseudoscience: Once we have learned something fundamental about the periodicity of heavenly bodies, for example, it is only a very natural extension to try to apply that knowledge to questions we have *not* been able to answer, the questions of human behavior. In other words, the oldest science, astronomy, generates the oldest pseudoscience, astrology, by the simple process of applying what we have learned about the universe to the questions we can't yet answer about human behavior: Likewise the science of chemistry validates the pseudoscience of alchemy. We use what we *do* know to explain what we *don't* know.

What we witness in eugenics is a simple extension of this principle. When Darwinism emerged, it was applied to human behavior by Galton (and independently by Spencer and others). But the application of an advance in science is simply a means of validating the social program that actually preceded it. Galton's program differs little at root from Gobineau's or from any other social tract of the 19th century that saw the wrong people proliferating and the destiny of civilization localized in their constitutions. Where the programs differed was in Galton's framing his work with Darwinism. In the next generation of eugenicists, the program remained the same, but the work was framed by Mendelism. The point is clear to all readers: Those old things you always thought were true about your own noble heritage and your superiority to others are now proven by the latest advances in science!

AMERICAN EUGENICS: THE PERIL OF THE HUDDLED MASSES

The leading American exponent of eugenics was Charles B. Davenport, Harvard-educated and well-funded by private foundations; his 1911 *Heredity in Relation to Eugenics* was a major early work that helped establish eugenics as a scientific program in America. Davenport's work is very instructive as a frank demonstration of the ways in which scientific ideas could be manipulated to lend credence to a set of social values.

First, he lays out the goals of eugenics, which are somewhat naive, to be sure:

> The general program of the eugenicist is clear—it is to improve the race by inducing young people to make a more reasonable selection of marriage mates; to fall in love intelligently. It also includes the control by the state of the propagation of the mentally incompetent. It does not imply the destruction of the unfit either before or after birth.[10]

Davenport's naivete lies first, fairly obviously, in the hope that he could convince anybody "to fall in love intelligently." Indeed, according to a historian of the eugenics movement, Davenport had trouble even convincing his own daughter.[11]

The second bit of naivete in the quotation, however, consists in the Faustian bargain of involving the government in social tinkering of the sort envisioned by the academic eugenicists. It lies in appreciating that once the state has decided that some people have intrinsic qualities that are best not passed into the next generation, it is simply more expedient—easier and cheaper—to kill them than to operate on them. And since the state is under constant pressure to trim its expenditures

Figure 5.1. Charles Davenport.

and spend those tax dollars (or Deutschmarks) wisely, the relative merits of birth control versus death control become a great deal fuzzier than Davenport recognized.

The question that immediately comes to mind, once we agree that certain qualities should not be passed on, is: What are those qualities? Here we can see the flaw of eugenics at its most obvious, namely, the arbitrariness of the traits it wishes to promote or limit. Davenport, for example, worried about the effects of syphilis on the populace:

> Venereal diseases are disgenic agents of the first magnitude and of growing importance. The danger of acquiring them should be known to all young men. Society might well demand that before a marriage license is issued the man should present a certificate, from a reputable physician, of freedom from them. Fortunately, nature protects most of her best blood from these diseases; for the acts that lead to them are repugnant to strictly normal persons; and the sober-minded young women who have had a fair opportunity to make a selection of a consort are not attracted by the kind of men who are most prone to sex-immorality.[12]

Thus, as Davenport is known to have been something of a prude, even by the standards of his own day, we can read him as declaring any sexually active person "abnormal".[13] One could easily wonder how Davenport's eugenics program would have emerged if he had been a vegetarian: Would he have declared all those who eat hamburgers abnormal and fit for sterilization?

Davenport and the eugenics movement circumvented the problem of arbitrariness and subjectivity with a brilliant construction: feeblemindedness. Feeblemindedness encompassed any mental defect, be it social, behavioral, or intellectual, and (since it was a phenotype) was easily diagnosable.[14] In this nebulous term, the eugenicists could isolate all forms of "abnormal" behavior, and then focus discussion on whether it was genetic or environmental in origin. Thus, in one infamous study purporting to demonstrate the heredity of feeblemindedness, Henry Goddard described some of his subjects in *The Kallikak Family*:

> The father, a strong, healthy, broad-shouldered man, was sitting helplessly in a corner. The mother, a pretty woman still, with remnants of ragged garments drawn about her, sat in a chair, the picture of despondency. Three children, scantily clad and with shoes that would barely hold together, stood about with drooping jaws and the unmistakable look of the feeble-minded.[15]

Since feeblemindedness was an unmistakable phenotype, it made the next questions sensible: Is feeblemindedness inborn? Is it a Mendelian unit character? Goddard answered the first question with respect to his good- and bad-blooded Kallikak lines bombastically:

> They were feeble-minded, and no amount of education or good environment can change a feeble-minded individual into a normal one, any more than it can change a red-haired stock into a black-haired stock. . . . Clearly it was not environment that has made that good family. They made their environment; and their own good blood, with the good blood in the families into which they married, told.[16]

And Davenport answered the second question readily in the pages of the journal *Science*:

> It appears probable, from extensive pedigrees that have been analyzed, that feeble-mindedness of the middle and higher grades is inherited as a simple recessive, or approximately so. It follows that two parents who are feeble-minded shall have only feeble-minded children and this is what is empirically found.[17]

We are thus presented with a hard, heritable characteristic encoded by a single gene, easily diagnosed, as the cause of social deviance—from crime to sexuality to poverty. It should be fairly easy to see that this implies a scientific program for the amelioration of social problems. For example, though admitting that "[w]e must . . . take all genetic studies of feeble-mindedness with a grain of salt,"[18] H. H. Newman of the University of Chicago proceeded to repeat the program but without criticizing it:

> Goddard and others maintain that there is a very intimate relation between crime, vice, and feeble-mindedness. Wipe out the feeble-mindedness, say they, and you wipe out most of the vice and crime.
>
> Feeble-mindedness has come to be the most pressing of all eugenic problems—one that should at once be recognized and solved if possible. Statistics seem to indicate that this defect is on the increase; certainly it is far too common to be ignored. . . . It has been estimated by one expert that in the United States one person in every 294 is feeble-minded; by another expert, one in every 138. . . . Calculations indicate that in the United States as a whole there are not less than half a million feeble-minded individuals, and several times that many individuals phenotypically normal but carrying the gene or genes for feeble-mindedness. A large proportion of these individuals are charges of the various states and cost the public many millions of dollars annually without contributing anything of value to the community.[19]

The next line goes unarticulated, but is strongly implied: Wouldn't we be better off if the feebleminded no longer existed?

Logically it stood to reason that feeblemindedness, as an inborn deformity condemning its bearer to a lifetime of misery, and condemning society to pay for it, could be curbed in two ways. The first would be to screen the people who were a burden on society and deal with them subsequently; the second, to prevent any more people so afflicted from entering society. Thus in a textbook on genetics published in 1913, Herbert E. Walter wrote:

> It is not enough to lift the eyelid of a prospective parent of American citizens to discover whether he has some kind of an eye-disease or to count the contents of his purse to see if he can pay his own way. The official ought to know if eye-disease runs in the immigrant's family and whether he comes from a race of people which, through chronic shiftlessness or lack of initiative, have always carried light purses. . . .
>
> The national expense of such a program of genealogical inspection would be far less than the maintenance of introduced defectives, in fact it would greatly decrease the number of defectives in the country. At the present time this country is spending over one hundred million dollars a year on defectives alone, and each year sees this amount increased.
>
> The United States Department of Agriculture already has field agents scouring every land for desirable animals and plants to introduce into this country, as well as stringent laws to prevent the importation of dangerous weeds, parasites, and organisms of various kinds. Is the inspection and supervision of human blood less important?[20]

There was another piece to the puzzle, however. Those defective peoples were not scattered across the globe at random. Feebleminded peoples seemed to be most prevalent among the world's populations *not* located in or derived from northern Europe. And it was Madison Grant, in *The Passing of the Great Race*, who assembled the full-blown eugenics platform, incorporating the Nordicism of Gobineau and the breeding program of Davenport, along with a calculus for emptying the jails and balancing the budget. His words at the time of the First World War have a sobering effect when one reflects upon the Second:

> A rigid system of selection through the elimination of those who are weak or unfit—in other words, social failures—would solve the whole question in one hundred years, as well as enable us to get rid of the undesirables who crowd our jails, hospitals, and insane asylums. The individual himself can be nourished, educated, and protected by the community during his lifetime, but the state through sterilization must see to it that his line stops with him, or else future generations will be cursed with an ever increasing load of misguided sentimentalism. This is a practical, merciful, and inevitable solution of the whole problem, and can be applied to an ever widening circle of social discards, beginning always with the criminal, the diseased, and the insane, and extending gradually to types which

may be called weaklings rather than defectives, and perhaps ultimately to worthless race types.[21]

The words, however, are those of an American, and an influential American (bearing the names of two Presidents), to boot. While Grant was himself a dilettante, his book carried a glowing preface by his friend, the leading evolutionary biologist of the generation, Henry Fairfield Osborn of the American Museum of Natural History. Here we find the eugenics program at its most lurid, with the validation of modern science, in the form that would be put into action by the German National Socialists: (1) Human groups are of unequal worth; (2) the difference in their relative value is constitutional; (3) the constitutionally defective groups should be kept out; (4) those which are already in require other measures.

In some cases, a work ostensibly on scientific matters of eugenics would degenerate into a diatribe against foreigners. In the case of *Racial Hygiene* by Indiana University bacteriologist Thurman Rice, the genetical science discussed in the rest of the book is simply forsaken in favor of a pompous xenophobia seemingly derived from the science. But the bottom line is always the same, having to do with restricting the input of alien elements from southern Europe into America:

In early days there came the English, the German, Swede, Welsh, Irish, Scotch, Dutch, and related peoples, and while these related stocks were coming the "melting pot" was a reality. It was an easy matter to fuse these people biologically; their customs were at least similar; there were no intense racial prejudices to overcome; their ideals were already essentially American; they were able to understand one another; they were homemakers and land-owners; they believed in education and democratic government, in law, order and religion. . . .

To-day the man who believes that the so-called "melting pot" will fuse the heterogenous mass dumped from the corners of the earth, in defiance of all laws of biology and sociology, into a desirable national type is either utterly ignorant of all the laws of Nature or is laboring under a most extraordinary delusion. . . .

We formerly received practically all of our immigrants from northern Europe. They were for the most part of an excellent type and would blend well together. . . . The situation is very different to-day; most of the recent immigrants who are coming to-day, or at least before the present law was passed, have come from eastern and southern Europe, and from other lands even less closely related; they do not mix with our stock in the "melting pot," and if they do cross with us their dominant traits submerge our native recessive traits; they are often radicals and anarchists causing no end of trouble; they have very low standards of living; they disturb the labor problems of the day; they are tremendously prolific.[22]

EUGENICS: SCIENCE AND PSEUDOSCIENCE

The difficulty in evaluating the eugenics movement in retrospect is that because it is such an extreme embarrassment to American biological science, there is a strong tendency to ignore it, deny it, or revise it. Eugenics was, in fact, a mainstream movement in the scientific community, cross-cutting political lines in its utopian vision of a crime-free society. Virtually all members of the genetics community were in favor of eugenics through the mid-1920s.

It is a consequence of the movement's popularity within the scientific community that eugenics was *science, not pseudoscience*. If all the relevant scientists believed it, how could eugenics possibly be pseudoscience? If eugenics represented a corruption of certain scientific principles, it is hard to escape the conclusion, from simply examining the literature, that it was the scientists themselves who were the corruptors.

> Man is neither angel nor beast; and the misfortune is that he who would act the angel acts the beast.
>
> —Blaise Pascal

Kenneth Ludmerer, a historian of the movement, notes that of the founding members of the editorial board of the journal *Genetics* in 1916, every one was a supporter of eugenics. Indeed, "until the mid-1920's no geneticist of note . . . publicly disputed [the claims of eugenics]."[23]

One of the earliest notable biologists to fall away from eugenics was Columbia's Thomas Hunt Morgan. While Princeton's E. G. Conklin was laying out the platform of eugenics without the evangelical zeal of other scientists in *Heredity and Environment*, he was nevertheless thoroughly uncritical of its central assumptions of racial rankings, immigration restriction, and feeblemindedness. So, too, Harvard's Edward East and William Castle.[24] Morgan, on the other hand, began to express his fundamental doubts publicly in a 1924 paper. First he noted the dual inheritance system (biological and social) operating in humans, and speculated that "our familiarity with the process of social inheritance is responsible, in part, for a widespread inclination to accept uncritically every claim that is advanced as furnishing evidence that bodily and mental changes are also transmitted." Next, he called upon the study of human heredity to become truly interdisciplinary: "competent specialists are needed to push forward scientific investigation— since other methods have signally failed." For, finally, when these other disciplines become actively involved in the nascent field of human heredity, "I believe that they will not much longer leave their problems in the hands of amateurs and alarmists, whose stock in trade is to gain notoriety by an appeal to human fears and prejudices—an appeal to the worst and not to the best sides of our nature."[25]

Alas, it was hard to tell the ideas of "amateurs and alarmists" from those of the professional biologists. After all, they served side by side on the advisory board of the American Eugenics Society.

In Morgan's 1925 *Evolution and Genetics*, he began to express doubts about the whole enterprise:

> The case most often quoted is feeble-mindedness that has been said to be inherited as a Mendelian recessive, but until some more satisfactory definition can be given as to where feeble-mindedness begins and ends, . . . it is extravagant to pretend to claim that there is a single Mendelian factor for this condition.[26]

> But it is not so much the physically defective that appeal to [eugenicists'] sympathies as the "morally" deficient and this is supposed to apply to mental traits rather than to physical characters. Ruthless genetic (?) [sic] reform here might seem too drastic and might be retroactive if pressed too far. Social reforms might, perhaps, more quickly and efficiently get at the root of a part of the trouble, and until we know how much the environment is responsible for, I am inclined to think that the student of human heredity will do well to recommend more enlightenment on the social causes of deficiencies rather than more elimination in the present deplorable state of our ignorance as to the causes of mental differences.[27]

> A little goodwill might seem more fitting in treating these complicated questions than the attitude adopted by some of the modern race-propagandists.[28]

Critics of the eugenics movement had been scarce before this time. In 1916, two scathing articles by the prominent anthropologists Franz Boas and Alfred Kroeber outlined the weaknesses and exposed the nonscientific nature of the eugenics movement. The articles do not appear to have had much of an impact upon the biology community, however, among whom public support for the movement did not begin to wane for about a decade.[29] Privately, several prominent biologists appear to have had their doubts as early as 1923, but shortly after Morgan's publicly aired skepticism, Raymond Pearl and Herbert Spencer Jennings, both of Johns Hopkins, began to criticize the movement in print.[30]

By this time, however, Congress had already passed legislation to restrict immigration of genetically undesirable populations into the United States. The Emergency Act of 1921 reduced immigration as a temporary expedient, with debate focused on an industrial labor glut. For the next three years, the focus shifted toward using the biological inferiority argument to justify the restriction of immigration, through the Johnson Act (named for Representative Albert Johnson, Chairman of the House Committee on Immigration and Naturalization). Since scientific ideas were known to support the prejudices on which the bill was founded, most of the debate invoked the published work of the eugenicists, several of whom were called to testify. Herbert Spencer Jennings gave a brief testimony undermining some of the most strident Nordic-eugenic claims. But he was clearly perceived as a minority voice within the scientific community. The only notable voice of science was the one Congress was hearing, and it was loud:

> While the Johnson bill was being debated, assertions by eugenicists were not being countered by persons of authority within genetics. Of the thousands of letters received by Johnson while his measure was pending, not one was from a geneticist or biologist, though several dozen were from important eugenicists.[31]

There was little reason to expect to hear contrariwise from biologists, since the eugenics ideas were not antithetical to their own. The effect of the Johnson bill was to cut back immigration of peoples from central, southern, and eastern Europe. Over the next decade this had the deeper effect of preventing the escape of many people who were ultimately exterminated by the Nazis in their eugenic fervor.

EUGENICS IN NATIONAL SOCIALIST GERMANY

The greatest mistake we can make in analyzing the eugenics movement retrospectively is to blame it on the Nazis. Certainly eugenics validated Nazism, as it validated other forms of racism and intolerance. But the Nazis merely implemented those ideas; they didn't dream them up.

When Adolf Hitler was writing *Mein Kampf* in prison in the early 1920s, he derived biological support for his views from a major textbook by three leading German biologists: Erwin Baur, Eugen Fischer, and Fritz Lenz.[32] That text was published in English in 1931 as *Human Heredity*, and though not specifically about eugenics, it concludes in a tone not unlike contemporary works from America:

> If we continue to squander [our] biological mental heritage as we have been squandering it during the last few decades, it will not be many generations before we cease to be the superiors of the Mongols. Our ethnological studies must lead us, not to arrogance, but to action—to eugenics.[33]

As an historical sidelight, Baur died in 1933; Lenz joined the Nazi party in 1937 as a department head in the Kaiser Wilhelm Institute for Anthropology; and Fischer joined the party in 1940, as director of the institute.[34] And yet their scientific views differed little from those of American geneticists in the late 1920s. Indeed, Fischer had been Davenport's personal choice in 1932 to succeed him as president of the International Federation of Eugenic Organizations.[35]

WHY EUGENICS FAILED

It is always easy in retrospect to see why a plan failed, or why archaic ideas were wrong. It is far more difficult to make judgments or predictions on the spot; yet when new ideas on human biology are raised, they generally require such evaluations. We can now see three major theoretical flaws in the eugenics movement of the 1920s.

First, there is the problem of reification. The declaration that social problems are attributable to feeblemindedness, and can thereby be bred out of the species, carries with it the assumption that feeblemindedness is a unitary entity. The fact that it can be given a single name, however, does not mean that it is a single thing. Indeed, taking the term in its most literal sense, we now know that mental retardation is very genetically heterogeneous. Any attempt to breed it out would therefore be very unlikely to succeed. And cutting back the cases of retardation caused by chromosomal imbalances would require tests *in utero*, not better matings.

The second problem is arbitrariness: While many would agree that it is a callous parent indeed who would knowingly and willingly pass on a serious genetic liability to a child, it is not at all clear exactly what a "serious" defect would be. Or more precisely, the decision on where to "draw the line" between a genetic trait acceptable for propagation and one unacceptable for propagation is a difficult one to make. While some might feel comfortable about placing the decision in the hands of an enlightened government, Clarence Darrow found it absurd to imagine that an institution that most people acknowledge to be inefficient, if not corrupt, could be relied upon to make wise decisions about who should reproduce. It would mean "that breeding would be controlled for the use and purpose of the powerful and unintelligent. . . . [I]t would bring in an era of universal sexual bootlegging."[36]

Further, it was clear from many of the writings that the qualities the eugenicists hoped to stamp out were not simply violence and mental illness, but also contrasting moral codes. While one may certainly speculate on the origins of different systems of values, and the problems that ensue for a society in which a significant proportion of the population has different standards of behavior, it is certainly unrealistic to proclaim that those people are irredeemably corrupt of germ-plasm.

The third flaw is hereditarianism, which piggybacks on the science of genetics, but is far older. The fact that many people pass on standards

ot behavior to their children that may be different from those passed on
to the children of the most affluent classes, is not apparently attributable
to genetics. To a large extent, therefore, the eugenicists were trying to
solve non-genetic problems through genetic means. In other words,
given that great strides had just been made in genetics, the program of
the eugenicists involved applying those advances fairly recklessly—by
taking an outstanding question (why different groups of people act differently) and applying an answer that happened to be appropriate to a
different question. Though we can identify feeblemindedness with mental retardation and discuss its heterogeneous genetic basis, the eugenicists used the term in a far broader sense—to encompass any behavioral
deviation from essentially middle-class standards.[37] Such a broad net
lumps together not just phenotypes that are genetically heterogeneous,
but those which may have no basis in genetics whatsoever. This enabled
the ancient hereditarian social values to derive scientific legitimacy from
genetics.

In retrospect as well, we can now see three important *practical* flaws
with the eugenics program. As a biological solution to social problems,
eugenics was looking for answers in the wrong places. Social problems
are caused by social circumstances and history, after all, not by genetics.
That the eugenics movement died out in America with the onset of the
Great Depression is probably no coincidence: as formerly wealthy and
powerful people joined the ranks of the impoverished and needy, it was
no longer possible to blame their situation on heredity. The "genes for
feeblemindedness" were simply overrun by economic forces, acknowledged geneticist Hermann J. Muller, as he fell away from the eugenics
movement.[38] Social problems, stemming from social causes, invariably
require social solutions.

Empirically as well, a breeding program would be doomed to failure
if the advocates paid attention to the genetics of animal breeding. As a
critic in *Scientific American* pointed out in 1932, "[t]he dairy cow, as a
cow, is not a very successful animal."[39] While bred for a particular feature, it nevertheless is not a hardy species, and in the wild would certainly fail to thrive. And yet the eugenicists made consistent analogies
to domesticated animal stocks in their appeal for more controlled
human breeding.

Further, pure-bred strains experience inbreeding depression, a condition named for a well-known loss of vigor that comes as a consequence
of homozygosity, or the loss of genetic variation. Consequently, pure-
bred strains have to be outbred to balance the appearance of the desired
trait with the vigor necessary for perpetuation of the strain; the result is
the opposite of inbreeding depression, or hybrid vigor. Yet the advocates
of eugenics sought a reduction in the biological variation in the species,

for in their assumption that a single type of organism is superior to others without qualification, they were forced to disparage genetic variation as a source of evolutionary novelty.

Though the genetics of animal breeding was a major focus of the Harvard eugenicist William Castle, it was his student Sewall Wright who developed a theory of evolution incorporating it.[40] Inspired by Wright, the geneticist Theodosius Dobzhansky developed a corpus of evolutionary theory constructed around the importance of genetic variability in a population, ultimately destroying the fundamental underpinnings of the genetic breeding program of eugenics. Thus, by 1937, Dobzhansky could casually dismiss eugenics in his classic *Genetics and the Origin of Species*:

> The eugenical Jeremiahs keep constantly before our eyes the nightmare of human populations accumulating recessive genes that produce pathological effects when homozygous. These prophets of doom seem to be unaware of the fact that wild species in the state of nature fare in this respect no better than man does with all the artificiality of his surroundings, and yet life has not come to an end on this planet. The eschatological cries proclaiming the failure of natural selection to operate in human populations have more to do with political beliefs than with scientific findings.
>
> Looked at from another angle, the accumulation of germinal changes in the population genotypes is, in the long run, a necessity if the species is to preserve its evolutionary plasticity.[41]

The last conceptual flaw is the primitive theory of history, or culture change, at the heart of the eugenics program. As a cultural theory generated by biologists, it suffered from the superficiality and dilettantism that bred it. The essayist H. L. Mencken was far more perceptive than the scientific community when in 1927 he criticized its premises about the processes of history. Mencken recognized that "superiority" is highly dependent on the time and place of birth:

> Before baseball was invented there were no Ty Cobbs and Babe Ruths; now they appear in an apparently endless series. Before the Wright brothers made their first flight there were no men skilled at aviation; now there are multitudes of highly competent experts. The eugenists forget that the same thing happens on the higher levels. Whenever the world has stood in absolute need of a genius he has appeared. . . .
>
> The eugenists constantly make the false assumption that a healthy degree of progress demands a large supply of first rate men. Here they succumb to the modern craze for mass production. Because a hundred policemen, or garbage men, or bootleggers are manifestly better than one,

they conclude absurdly that a hundred Beethovens would be better than one. But this is not true. The actual value of genius often lies in its singularity.[42]

Beethoven, of course, was not only the victim of physical infirmity, but, Mencken points out, "the grandson of a cook and the son of a drunkard."

To anthropological critics of eugenics, the earliest from within the scientific community, a major liability of the field was that the limiting factor in social "progress" is *not* the rate of production of geniuses, but the accessibility of resources to the pool of talented people who may be able to benefit from them. The less accessible the resources in a society, the fewer the number of people who can contribute productively to its "advancement"—regardless of the distribution of inborn talent among them. Social history is thus not an organic property of people, but a "superorganic" property of social and cultural systems, in the classic formulation of Kroeber's.[43]

LESSONS FOR OUR TIME

Eugenics represented a major failure on the part of mainstream American science to divorce human history from biology. One can certainly not fault thinkers for failing to be ahead of their times,[44] yet in the eugenics movement there is an ignorance that transcends simple unenlightenment. After all, the anthropologists by the 1920s had already drawn the conclusion that human biology and social history were separate classes of phenomena. Franz Boas particularly, in his widely read *The Mind of Primitive Man*, began by demonstrating that "historical events appear to have been much more potent in leading races to civilization than their faculty, and it follows that achievements of races do not warrant us in assuming that one race is more highly gifted than another."[45] Thus, the pronouncements of the eugenicists involved more than just making racist or hereditarian assumptions about human behavior and history—it involved ignoring or dismissing the conclusions of leading anthropologists on the matter.

The paradox is that by 1925, Clarence Darrow could publicly ridicule William Jennings Bryan during the Scopes trial for his lack of knowledge of science by asking: "Did you ever read a book on primitive man? Like Tylor's 'Primitive Culture' or Boas or any of the great authorities?"[46] Bryan had not. But the large-scale ignorance of anthropological knowledge was clearly not limited to creationists.

The biologists promoting eugenics had few or no reservations about representing their views as those of the scientific community. They were, indeed; but they were the views of a scientific community that neither knew nor cared about the judgments of the scientists actually working in the area of interest. It may be difficult to imagine the scientific community pronouncing on some issue of fruitfly biology in spite of what fruitfly biologists think about the matter, but such a scenario is crucial for understanding the scientific validation behind the eugenics movement. Scientists have social values *as citizens* that they bring to the study of human biology; they do not bring comparably obvious values to their study of flies or birds. Thus, speaking and writing outside areas of expertise, these eugenic scientists were little more than well-educated laymen, just as an expert on human biology and history who wrote a book on ornithology would be regarded.

Reviewing the reception of Madison Grant's *The Passing of the Great Race*, which he had reviewed favorably in *Science*, Harvard's Frederick Adams Woods summarized: "Nearly all the reviews published in scientific journals or in the leading newspapers were either favorable or moderately favorable."[47] Woods knew he was with the majority as he propagandized for "science" and against the social science that undermined the work.

The attraction of eugenics lay in its easy answers to complex problems, and in its idealism: ultimately the goal was to improve society. For this reason, it cut across political lines, being as attractive to liberals who wished to make the world different and better, as to conservatives whose own success was thereby scientifically validated. Only well after the movement had been widely criticized by people outside genetics and biology did the biologists begin to fall away from the movement. Possibly they were late to do so because the eugenics movement was advancing the cause of genetics and biology in America—which brought greater attention to the work biologists were doing, particularly during the era of the Scopes Trial, and greater funding potential.

In retrospect, such a bargain is clearly Faustian. If the biologists did in fact widely see the abuse to which genetic knowledge was being put, but refused to criticize it out of self-interest, they paid dearly for it. As historians of genetics have noted, the eugenics movement ultimately cast human genetics is such a disreputable light that its legitimate development was retarded for decades.[48]

However early on some biologists may have acknowledged the excesses of eugenics, one need only pick up any genetics or biology book from the early 1920s to see eugenics as a major topic, and see it discussed favorably. To the extent that mainstream geneticists had reservations about the movement up until the mid-1920s, they rarely expressed them.

A case in point is Sewall Wright, who like virtually all geneticists in America, was on the advisory board of the American Eugenics Society. He had studied under Davenport at Cold Spring Harbor, and Castle and East at Harvard. According to his biographer, Wright never published explicitly on the subject of eugenics, neither publicly advocating the ideals of eugenics nor publicly repudiating them. When asked by a leading eugenicist in 1932 to rebut T. Schwann Harding's *Scientific American* paper that had been critical of eugenics, Wright politely declined. On the other hand, "[a]s an idea for benefiting mankind, Wright had no theoretical objection to eugenics," and did not mind being on the American Eugenics Society's stationery, but had no active participation in the movement.[49]

And yet, Wright was a signatory to the 1926 report of the Committee on Research of the American Eugenics Society, which set forth as goals the study of "internal factors that contribute toward criminalistic reactions," "the consequences of particular matings, like those between north-western Europeans and Jews," and the "net increase of inferior stocks"—the mainstream science of eugenics.

Faced with the embarrassment of the field of biology promoting or tacitly condoning the eugenics movement, biologists have often treated the subject with some considerable degree of revisionism. One way to avoid responsibility for the biological community's involvement is to ignore eugenics completely, as geneticist Leslie Dunn did in his *A Short History of Genetics, 1864–1939*.[50] Another is to blame eugenics on the Nazis, who ultimately implemented it most starkly. Clearly, however, the ideas were far more widespread, pervasive, and scientifically mainstream than would be justified by either of these alternatives. A third approach is to imagine that most geneticists were actually opposed to the movement, but were led or duped by a small number of zealots. Thus, two modern writers lament the fact that "[Herbert Spencer] Jennings, who was a strong opponent of the eugenics movement," was given little time to speak at the congressional hearings on immigration in 1924.[51] He would presumably have laid the movement to rest with his testimony. And yet, that would leave unexplained why Jennings was offered the presidency of the American Eugenics Society in 1926! That he declined, to devote himself wholeheartedly to laboratory research,[52] is more understandable than why the society would want to be led by "a strong opponent" of its aims. Jennings in 1924 was beginning to have reservations about certain aspects of the eugenics program—but he was far from an opponent of the movement, and had been an enthusiastic supporter, like most other geneticists.

It is often hard to tell good science from bad science in any way but retrospectively. But since science, in the context of our cultural values, lends validity to ideas, scientists are ultimately responsible for the ideas promoted in its name. Particularly in the human sciences, where lives can be wasted or cultivated according to the ideas that are considered scientific, we require particular vigilance in distinguishing good from bad science. There is an immediacy to the judgments that must be made in the human sciences, for to make them in retrospect is to make them belatedly, when lives may be at stake. And when popular social prejudices are proven scientifically, they need to be scrutinized with particular care.

It would be a comforting thought to know that all the mistakes that could be made in the study of human variation have actually already been made. Then our task would be simply to try and make sure we do not repeat them.

NOTES

1. Kevles (1985).
2. Galton ([1869] 1979:1).
3. Ibid., p. 38.
4. Ibid., p. 338.
5. Ibid., p. 342.
6. Pearl (1927:263). Pearl's defection from the eugenics ranks was reported widely in the newspapers.
7. Box (1978), Mazumdar (1992).
8. The rancorous dispute in England between the Mendelians (led by William Bateson) and biometricians (led by Karl Pearson) influenced the nature of the eugenics movement. Pearson (e.g., 1909) was the leading spokesman for eugenics in England, which naturally took on an anti-Mendelian slant, against which Bateson (e.g., [1919] 1928) gravitated. In America, where Mendelism caught on largely unopposed, eugenics was almost wholly Mendelian in character (Ludmerer 1972:157).
9. White (1947), reprinted in White (1949).
10. Davenport (1911:4).
11. Kevles (1985:52).
12. Davenport (1911:2).
13. Literally, of course, any sexually active *man*. Davenport takes for granted the Victorian stereotype that women are sexually passive.
14. The term had come into use in the 1870s, and by the first decade of the twentieth century was a common catchall term for people with psychological or social problems, before being appropriated by the eugenicists. See Trent (1994).
15. Goddard (1912:77).
16. Ibid., p. 53.
17. Davenport (1921:393).
18. Newman (1932:459).

19. Ibid., pp. 460–61).
20. Walter ([1913] 1932:522).
21. Grant (1916:46–47).
22. Rice (1929:301–2). The reviews for *Journal of Heredity* judged the book to be "really excellent" (Anonymous 1930).
23. Ludmerer (1972:25). The founding editorial board in 1916 consisted of George H. Schull, William A. Castle, Edwin G. Conklin, Charles B. Davenport, Bradley M. Davis, Edward M. East, Rollins A. Emerson, Herbert Spencer Jennings, Thomas Hunt Morgan, and Raymond Pearl. Though several of these became disaffected from the eugenics movement by the mid-to-late 1920s (notably Morgan, Jennings, and Pearl), others were consistent in their advocacy (notably Castle, Conklin, East, and of course Davenport himself).
24. Conklin (1922), East (1927), Castle (1930a).
25. Morgan (1924:408–9).
26. Morgan (1925:200–1).
27. Ibid., p. 205.
28. Ibid., p. 207.
29. Boas (1916), Kroeber (1916). Though not widely cited by geneticists, the thoughts of Boas on eugenics may have had some indirect effect. Thomas Hunt Morgan resisted associating himself with the American Eugenics Society, and was the first American geneticist to challenge the movement's ideas. While it is certainly possible that Morgan was simply more insightful than his contemporaries in this regard, it is also true that he and Boas both worked in Columbia's Schermerhorn Hall.
30. Jennings (1925), Pearl (1927). The English eugenics movement also began to be criticized primarily by the Marxists Lancelot Hogben and J. B. S. Haldane (Barkan, 1992).
31. Ludmerer (1972:113).
32. Müller-Hill (1988:8).
33. Baur et al. (1931:699).
34. Müller-Hill (1988), Proctor (1988). See also the damning discussion of Eugen Fischer as a craven opportunist by Goldschmidt (1942), who had been forced to leave his position of director of the Kaiser Wilhelm Institute of Biology on account of his ancestry.
35. Müller-Hill (1988:9). On the intimate relationship between American genetics of the 1920s and German genetics of the 1930s, see Kühl (1994).
36. Darrow (1926:137).
37. Trent (1994).
38. Muller (1933).
39. Harding (1932:25).
40. Wright (1986).
41. Dobzhansky (1937:126).
42. Mencken (1927).
43. Kroeber (1917), following Herbert Spencer's terminology.
44. In the words of anthropologist Clifford Geertz (1988, p. 50), "let him who writes free of his time's imaginings cast the first stone."
45. Boas (1911:17).

46. *New York Times,* July 21, 1925; Ginger (1968:171).
47. Woods (1923:95).
48. See especially Kevles (1985).
49. Provine (1986:180, 182).
50. This is a particularly instructive example, since the 1925 edition of Dunn's popular textbook of genetics ended with a very conventional chapter advancing the case of eugenics; but by the 1938 third edition the word did not even appear in the index. In the 1930s, and especially after World War II, Dunn became an outspoken opponent of scientific racism. See Marks (1993).
51. Garver and Garver (1991:1110).
52. Ludmerer (1972:81).

6

Racial and Racist Anthropology

The existence of between-group differences in the human species has been approached in several different ways. Racist anthropology is one in which group properties are attributed to individuals, and their worth judged accordingly. This violates biological principles and stands in opposition to our ethical values. Racial anthropology seeks to study biological differences among human groups. This study has been done using many criteria, but often the questions are put in the Linnaean framework: How many races are there and what are they?

RACISM AND EUGENICS

Though it may seem strange by contemporary standards, there was a difference (in spite of considerable overlap) between the positions of the eugenicists and racists of the 1920s. Where eugenicists wanted to breed a better form of citizen, racists maintained that certain forms of humanity were constitutionally superior overall to others.

It is not hard to see how such positions could be complementary. If certain groups were uniformly superior to others on a constitutional, or biological, basis, then it stood to reason that those groups should be encouraged to proliferate. Thus racism and eugenics had a broad common ground.

Yet by the 1920s a slightly more liberal mode of thought had developed among scientists interested in human variation. If there were constitutional differences among human groups, but these varied in a non-uniform manner—one group being better at certain things, another group being better at other things—then the basic claim of racism, that groups are "superior" or "inferior" to one another *overall*, would be undermined.

The concession is minor by modern standards, but was a significant break in the apparently united "scientific" view of human differences. The leading student of human diversity, Harvard's Earnest Albert Hooton, could maintain on the one hand that races were not linearly rankable, and on the other hand, that we still needed to breed a better form of citizen:

> Each [race] has, in all probability, its own array of points of strength, off-set by weaknesses; and these points do not always coincide in all of the different races. Add them all together in any single race, and I am afraid that it amounts to zero—or, in other words, it comes out even. Thus, all races are equal.[1]

> I believe that this nation requires a biological purge if it is to check the growing numbers of the physically inferior, the mentally ineffective and the anti-social. These elements which make for social disintegration are drawn from no one race or ethnic stock. Let each of us, Nordic or Negro, Aryan or Semite, Daughter of the Revolution or Son of St. Patrick, pluck the beam from his own eye, before he attempts to remove the mote from that of his brother. Every tree that bears bad fruit should be cut down and cast into the fire. Whether that tree is an indigenous growth or a trans-plantation from an alien soil, matters not one whit, so long as it is rotten.[2]

While this is still a rather inhumane view of the causes of and remedies for social problems, it differs from that of contemporary racists by rec-ognizing that favorable traits are heterogeneous and are not uniformly distributed across human groups, and in identifying the undesirable individuals fairly eclectically.

Likewise, geneticist William Castle, whose views on the subject were strongly influenced by his Harvard colleague Hooton, could deplore "[w]riters who appeal to race prejudice[...] that our group of races is the best group, our particular race the best race and all others inferior,"[3] but at the same time:

> Consider for a moment the physical (not social) consequences in the United States of a cross between African black races and European whites, an experiment which has been made on a considerable scale. The white race has less skin pigment and more intelligence. The first difference will not be disputed, the second can be claimed at least on the basis of past racial accomplishment.[4]

In this meaty quote, Castle reiterates the false inference of individual abilities from social history as his illustration of the intellectual caliber of black people. But it could at least be consistent with his earlier chas-tisement of racists if "intellect" is seen as just a single character, and not an overarching determinant of racial value. Of course, many eugenicists (following Davenport and Goddard) did indeed see intellect as the major determinant of the worth of an individual, and consequently of groups as well—hence their emphasis on establishing the genetic basis of "feeblemindedness".

HUMAN DIVERSITY

The attitude adopted by Hooton, which seems slightly paradoxical in some respects and certainly unenlightened by modern standards, nevertheless subtly incorporates a major advance in the study of human variation. By maintaining that there are people of all races with favorable qualities and people of all races with unfavorable qualities, Hooton made a major break with the racist eugenicists, who wanted whole ethnic groups sterilized or otherwise ostracized on account of their undesirability. By focusing attention at the level of the individual, rather than the

Figure 6.1. Earnest Hooton.

population, Hooton drew attention to *polymorphism*, the biological variation that exists within populations.

The study of race is necessarily the study of differences among groups of people. But the relation of the differences that exist *among* groups to the variation that exists *within* human groups was uncharted territory at the time. As we will see in the next chapter, the application of genetic techniques to these questions showed that within-group variation (polymorphism) indeed held the key to understanding biological diversity in the human species. But as long as the focus of the science of human diversity was on "race", polymorphism would generally be ignored. The emphasis would be on polytypism, the study of the differences among groups, which was the reason for pursuing the study of race in the first place. Hooton's insistence on the importance of polymorphism in human populations was thus a crucial advance, though not widely appreciated at the time.

This gives us, however, a chance to make a critical distinction in the study of human diversity: the distinction between racial studies and racist studies. Racial studies are examinations of the biological differences among human groups. To the extent that such differences exist, they should be subject to documentation and analysis as dispassionately as one might go about documenting the differences among populations of clams.

The study of human diversity, however, has often been carried out with an implicit value judgment—something that does not form a part of the study of biological diversity of other species. In humans, the biological differences among human groups reinforce the social divisions that may also exist. If all social groups received equal treatment—had equal rights and equal opportunities for advancement—the study of the biological differences among them would be straightforward. It is not so straightforward, however, since the differential treatment often accorded to different groups can find a validation in the biological differences that may accompany them.

In other words, locating constitutional differences among human groups can provide a justification for treating those groups differently. This presents students of human diversity with an ethical dilemma not faced by students of clam diversity: although nobody really desires to treat clams differently from one other, the official scientific designation of *human* races can affect people's welfare. It therefore places a burden of responsibility on the scientist not only to ensure that the results are derived with the highest possible degree of rigor (a general aspect of competent scientific work), but to monitor the application of those results. The study of human biological diversity is not value-neutral, like the study of clam diversity: it can affect the quality of people's lives, and the scientist ultimately shares responsibility for it. One of the major differences between the study of human variation a few decades ago and the present is the acknowledgment of responsibility for one's scientific conclusions.

Not only are people's lives affected as subjects of racial analysis, but because they know it, people are often very concerned about *how* they are classified by the scientist—again, a problem the taxonomist of clams does not face. When the Armenian community of Washington state contacted Franz Boas in the 1920s over their racial status, it was more than a purely intellectual exercise. Because Armenians had been measured to be brachycephalic (broad-headed) like Asians, they were classified as non-Caucasian and on that basis were denied the right to own property. Boas gave expert testimony on the plasticity of head form, and invalidated the scientific basis for this racial classification.[5]

RACIST STUDIES

Human variation comes in packages of two sizes: the individual and the group. In a racial study, the emphasis is on the analysis of biological differences between human groups. To what extent might this illuminate the biology of the individual, who is both a part of a population and, at the same time, an autonomous biological and legal entity? Here is where we find the intellectual baggage that has been carried by studies of human diversity for centuries.

We define a racist study as one in which the individual is judged on the basis of group membership, and the qualities attributed to the group are therefore considered to be represented in the individual. It involves subsuming the biology of the individual to that of the group to which it belongs (or is attributed). The logic of racism is shown in the following syllogism:

Scots are frugal.
You are a Scot.
Therefore, you are frugal.

In this context it is irrelevant whether or not you are in fact frugal: you *must be* frugal by virtue of being Scottish. In other words, the empirical basis of knowledge has been discarded. One does not need to observe whether or not you are actually frugal; for by this train of thought, that attribute is yours simply by virtue of having been born Scottish.[6]

It is important to appreciate that the rejection of racist thought is a very recent development, and is a product of our own culture. There are three good reasons to reject racist thinking, all of which are culture-bound. First, it ignores or rejects empirical evidence. To the extent that the utility of a proposition ("you are frugal") is customarily determined by observation or evidence, the proposition as given is not useful, since observations of your own frugality, or lack of it, are simply considered irrelevant within this frame of reference. But by the standards of knowledge in our society, an assessment of your frugality can only be made by observing you, not by generalizing from observing others.

Second, generalizations about a group are notoriously difficult to validate and sustain. On what basis do we know that the Scots are indeed frugal in the first place? Perhaps it is simply an undeserved reputation, flattery or slander, as many such group generalizations turn out to be. If indeed the Scots are frugal, do we know how any particular Scot comes by it? Whether it is constitutional and instinctive, or learned and taught, may have very different implications for the proposition that a given Scot is or is not frugal. Is an unborn Scot destined for frugality? How can we find out? For example, do emigrants from Scotland and their descendants remain frugal? Do immigrants to Scotland and their descendants become frugal? Would a Scot raised by Danes be frugal, or not?

Thus, the basis for group generalization itself needs to be assessed, and also its significance as a predictor of individual behavior. Obviously, adequately controlled studies are virtually impossible for generalizations of this nature. Then how do we judge?

The burden of proof in science always falls upon the person making the claim. This is one of the major distinctions between science and pseudoscience: Pseudoscientists challenge others to spend time refuting their claims, while scientists gather evidence and assess the body of evidence that exists to validate their own claims. Therefore, without adequate demonstration of the generalization itself or of its intrinsic nature, a racist claim—that a person has by virtue of birth the presumed attributes of the group—is invalid until proven otherwise. Studies of the behavioral patterns of immigrants—studies of acculturation—show that human behavior is extremely plastic. We adopt some new ways from others, and they adopt some of ours; thus our own behaviors are different from those of our own recent ancestors, as well as from those of unrelated peoples. To suggest that some group behaviors are constitutionally rooted demands a strong burden of proof, rarely if ever met.

Finally, the rights of the individual are fundamental to our society. Most importantly, a person has the right to be judged *as an individual*,

> What *one* Christian does is his own responsibility, what *one* Jew does is thrown back at all Jews.
>
> —Anne Frank

and not simply as a group member. To judge persons not as they are, but as the group to which you attribute them may or may not be, goes not only against contemporary standards of rationality, but against the very foundations of our society.

Basic scientific racism surfaces in different forms, which all retain the common feature of judging an individual by presumed properties of the group. In one manifestation, the assertion that different races differ by virtue of certain ingrained or instinctive behaviors generally localizes the group attribute within the constitution of the individual. Likewise, judging different human groups on the basis of their levels of technological advancement (i.e., the confusion of social history and biology), also localizes group properties within the constitution of the individual. These assertions appear to be both inaccurate and racist.

But again, it must be reiterated that these criticisms are only valid in the cultural framework of a modern liberal democracy. The very idea of monarchy, or simply of a hereditary aristocracy, is founded on principles that we would identify as racist—the individual members of *this* group of people are born constitutionally superior to *that* group of people. Yet these are principles intrinsic to societies over most of the world throughout most of history. These racist principles, in fact, only began to be superseded by cultural changes in Western Europe and America in the last few centuries.

RACIAL STUDIES

Unfortunately, throughout the history of anthropology, it has proven difficult to distinguish between racist studies and racial studies. The reason is not hard to find: scientists themselves, educated and functioning within a cultural matrix, bring their own cultural values to whatever they study. There are relatively few cultural values one can bring to the study of fruitflies, relatively few ways in which the quality of a fly's life could be threatened by the values promoted by the scientist, and it is difficult for us to empathize with a fly whose quality of life was actually diminished by the pronouncements of scientists.

It is altogether different with humans. Though humans are more interesting than flies, the recognition that social values impinge upon the study of humans, and considerably less so upon flies, is a recent one. The first victim of that recognition was probably Carleton Coon (Chapter 2). Coon, a student of human variation who had advanced a poorly received theory of human evolution, was criticized by the evolutionary geneticist Theodosius Dobzhansky (Chapter 5), whose own primary research was on fruitflies. Dobzhansky chastised Coon for permitting his scientific work to be invoked by segregationists, and by others with oppressive political agendas. Coon replied:

> Dobzhansky states that "It is the duty of a scientist to prevent misuse and prostitution of his findings." I disagree with him. It is the duty of a scientist to do his work conscientiously and to the best of his ability . . . and to reject publicly only the writings of those persons who . . . have misquoted him. . . .
>
> Were the evolution of fruit flies a prime social and political issue, Dobzhansky might easily find himself in the same situation in which he and his followers have tried to place me.[7]

In retrospect we can marvel at Coon's naivete. To suggest that a theory about human biology and history is value-free, and that scientists can therefore be aloof and oblivious to the applications of their ideas, is absurdly archaic. And yet, he was right in one respect: His own theories were subject to greater scrutiny because they concerned humans specifically. Coon was faced with the responsibility of passing scientific judgment on human diversity, and failed to meet it. It ultimately destroyed his career.

Coon's mistakes (Chapter 2) were inferring race from fossils, using cultural criteria for ranking races, and ranking races on very poor evidence by inferring different times for becoming human. The premise, that there are large clusters of humans that differ fundamentally from other large clusters, was the basic premise of the study of human diversity for centuries. And yet, as scholars began to accept that they and their colleagues were responsible for the science of humans they produced, a new fundamentally threatening question began to surface: Why bother with racial studies at all, if they seem invariably to lead to racist conclusions?[8]

Racial studies need not imply racist conclusions. The connection, however, seems to be that the classes and institutions with the greatest interest in studying and in supporting the study of differences among human groups have traditionally been those to whom the identification of such differences is most significant. They are consequently those to whom the establishment of boundaries between the groups is also highly significant. Thus, the interests of racism and racial studies have often coincided.

WHAT DO DIFFERENCES AMONG HUMAN GROUPS REPRESENT?

Probably the most fundamental difficulty with racial studies in the first part of this century was the unclarity about what these basic divisions of the human species actually represented. To polygenists of the 18th and 19th centuries, they represented independent creative acts on the part of God, an interpretation that obviously served to reinforce the differences among the races, and to undermine any suggestion that they be entitled to equal rights.

With the triumph of Darwinism, the monogenist view was held virtually unanimously, which required accepting the differences among races to be due to evolutionary forces. But even if the source of the differences was now understood, it was still not clear just what the living races of humans in fact were. Were, they, for example, the remnants of the original, primordial, human groups, who then reproduced and expanded to fill large land masses? Such a position was just a slight modification of the polygenist position, and although implying common ancestry for all the races, it nevertheless also implied that the differences between them were ancient and profound. The task of the anthropologist, therefore, was simply to find where the "natural divisions" occurred among aboriginal populations of the world.

But this interpretation of race has theoretical as well as practical difficulties. On the practical side, the essentially continuous nature of human variation had been acknowledged, certainly, since Blumenbach; therefore "natural divisions" were all but precluded. As this approach to race seemed to have very little relation to contemporary populations of the world, it was soon necessary to acknowledge modern populations as "mixed races," the result of extensive hybridization between originally discrete and "pure" populations, resulting in the continuous distribution of features we now encounter. This, however, renders the study of race sterile, as it would have little application to the real world.[9]

By the early 20th century, racial studies had changed fundamentally.

A race was no longer a real series of populations, no longer a large cluster of real human groups, but an abstraction. It became, rather, a series of qualities or characteristics that were located in individual people. Thus, to Harvard's Roland B. Dixon (1923) populations were composed of races, rather than vice-versa—and he could thus identify the relatively few human races within many different aboriginal populations:

> [T]he vast majority of living men must have a complex racial ancestry, and such a thing as a pure race can hardly be expected to exist. However distinct, therefore, races may once have been, the peoples of the world to-day are complex mixtures of these original types, in which we must seek to discover, if we can, the constituent elements.[10]

Dixon used three sets of measurements: cranial index (head shape), altitudinal index (face shape), and nasal index (nose shape), and constructed races from the various permutations of these indices. Thus, his Alpine type contained a particular combination of features: brachycephaly, hypsicephaly, and leptorrhiny (round head, high head, narrow nose). He was able to locate this type of skull among the aborigines of Switzerland, Hawaii, and China, though in varying frequencies. Likewise, his proto-Australoid (long head, low head, wide nose) could be spread through Australia, Egypt, and California.

The concept of race had made an extraordinary reversal in the hands of the anthropologists of this era. Where race was supposed to be something equivalent to the subspecies—a large cluster of populations diagnosably distinct from others, though interfertile—it had now become something metaphysical—a category to which individuals would be allocated, regardless of the biological history of the population from which they were extracted.

If this seems strange, it is because of the relatively new recognition of variation, and the shift in emphasis from the archetypal mega-populations to the variable individuals within them, the implications of which anthropologists would not fully grasp for a few more decades. Thus, to Hooton, in about as definitive a statement as could be made by an anthropologist studying human variation, the challenge to the scientist was racial "diagnosis"—to discern from the complexities of a person's appearance their race. And it was tricky, because one could look white and really be black, and vice versa:

> I am of the opinion that racial characteristics are better defined in the skeleton than in the soft parts . . . Many individuals of mixed blood, who are fundamentally white, show characters of skin, pigmentation, and soft parts which would lead a superficial observer to classify them as pre-

dominantly negroid. But the skull and framework of the body may show a basically non-negroid morphology. On the other hand, it is equally true that some persons who appear to be white show definite negroid or mongoloid skeletal features.[11]

In other words, race was a presumably biological feature that an individual had, but was composed of disparate elements that might contradict one another within a single body. Nevertheless, races—whatever their nature—had to be defined, and real people had to be diagnosed, as a physician diagnoses an illness, through a close examination of the symptoms expressed. So what was a race? "[A] vague physical background, usually more or less obscured or overlaid by individual variations in single subjects, and realized best in a composite picture".[12]

Thus, very few persons could really be said to represent a race within this conception—the fact of variability within real human populations simply undermined the applicability of such an idea. So race was an abstraction, identifiable to varying degrees in people, but perfectly applicable to rather few of them. And what was the nature or source of this abstract racial type? Was it supposed to be the genotype of a distant progenitor? Hooton was never clear on just what the racial type represented. Mostly he seems to have regarded it as an artifact of statistics, an ideal Platonic image.[13]

To others, however, those races that came together in various manners in living humans represented prehistory: the migration of ancient populations. In this view, modern populations showing continuous intergrading variation are the result of gene flow between originally distinct and fairly homogeneous primeval races.[14] If you could trace ancestry back through time, you would find groups being less and less diverse, and ultimately encounter a small number of distinct homogeneous populations of humans. However, it was difficult to reconcile this view with reality—namely, that human groups have been migrating and interbreeding throughout the entire course of recorded history. There consequently seems no reason to think there was ever a time when they were *not* doing so.

These different concepts of race all shared an important assumption. If races are discrete groups of populations, the continuous nature of human variation undermines the utility of race as a basic way to study the species; and if they are ideal forms, their applicability to living groups of humans is somewhat dubious. But either way, race was taken to be something relatively stable. Race was something that was fundamental to human biology, probably nearly as old as the species itself. Consequently "race mixture" was usually taken to be a recent phenomenon, an inconvenience for the student of human variation.[15]

PERFORMANCE AND ABILITY

Racial and racist studies merged in one specific arena in the early 20th century: the detection of significant differences among groups of people in how they did on tests. The most famous of these were the IQ tests given to army recruits during World War I (in which blacks scored lower than whites), and tests given to immigrants (showing them to be inferior to the people who were already here). These attempts to locate group differences ended up by condemning members of groups on the basis of how other group members did on tests.[16]

Intelligence tests have, of course, changed drastically since those days, and the assumptions that go into their interpretation have changed as well. One false assumption surfaces in many different areas, however, which was most visible in the early IQ tests: The inference of ability from an observation of performance. This is often articulated in a different form— "heredity" versus "environment"—which shifts emphasis toward genetic etiology, and away from the basic flaw in deriving certain results from certain data.

Let us say that I encounter two differences between large groups of children: one group has had a greater success rate on algebra problems, and the other has had a greater success rate in making foul shots with a basketball. Let us say that this difference is consistent and replicable (a situation that is not terribly far-fetched). How do I interpret the group difference, and what am I reasonably entitled to think about the individual children composing the groups?

The data show that the group differences are real: though some children in one group may overlap some in the other group, the averages are nevertheless statistically different. But what might such differences imply? The key recognition here is that we have measured *performance*, a single event determined by many factors. But often we are trying to infer *ability*, or what that person is capable of, given optimal circumstances. Unfortunately, unless we actually optimize the circumstances, we have no real way of inferring one from the other. This is the same problem mentioned at the end of Chapter 1 in a slightly different context: the problem of inferring potentials.

In other words, we know that those who perform the algebra problems correctly, or make 10 free throws in a row, had the potential to do so—because they did it. But how do we know from the observation that they *didn't* do it, whether they *could have* done it? The relationship between the two categories, performance and ability, is highly asymmetrical. It is trivially easy to infer positive ability from a positive performance, but it is difficult to infer negative ability from a negative performance. To do so, we would need a comprehensive listing of the factors that might affect a performance—from endowments like eyesight and coordination, through simple variables of circumstance like financial and nutritional status, to complex developmental factors like parental attention, value systems, self-image, and aspirations. Performance is contingent upon many things, only one of which is ability. In fact, given the extraordinary complexity of the factors that can affect performance across groups on any set of tasks, it may well be surprising that there is such uniformity of results!

Nevertheless, the problem is clear: to infer differences in ability from differences in performance, one needs to control for an incredibly diverse array of factors. One generalization has been clear throughout most of the latter part of this century: the more variables are controlled, the more similar the performance of two groups on any series of tests. It would appear that human groups have roughly equivalent abilities.

Let us make this generalization very clear, however. That human beings differ from one another in their hereditary qualities has never been seriously questioned. At issue are (1) the nature of the differences encompassed by such a statement, and (2) the nature of the variation among *groups* of humans. When we say that human groups do not appear to differ in their abilities, we are *not* saying that their performances are identical; we are *not* saying that people do not vary in their native abilities; and we are *not* saying that there are no biological differences among human groups. We are saying that: (1) performances are not adequate measures of abilities; (2) people vary from one to another in their abilities (for the genotype is a very personal thing) but such variation is not translatable to the variation between groups, each of which contains a range of people with varying abilities; and (3) human groups differ biologically from one another (for example, in appearance), but not convincingly in the complex genetic factors (whatever they may be) that compose general abilities.

RACE AS A SOCIAL CONSTRUCT

Having articulated the problem, we are now in a position to sidestep it, as an earlier generation was not. We may observe differences in performance between groups without inferring group differences in abilities; we may acknowledge the continuous nature of human variation, and the variation of individuals within a population. What, then, can we say about the study of race in the human species?

The answer requires one last recognition from the work of Hooton, which became visible to the intellectual generation that succeeded him. Hooton was concerned with identifying a person's race, which could be particularly difficult in people of mixed ancestry. Yet this contains a genetic paradox: most people will be assigned to one race or another, but are acknowledged to have variable amounts of hereditary participation in several. How can these facts be reconciled?

The solution has only been appreciated in the last few years: Race is largely a social category, not a biological category. The heredity of race correlates to some extent with genetics, but is principally derived from a non-scientific, or folk concept of heredity. Consider the offspring of a union between one person from central Africa and one from western Europe. What is the race of the children?

It should be obvious that genetics precludes a simple answer: The children participate equally in the ancestry of both parents, and consequently represent the gene pools of both populations. And yet, the children will almost invariably identify themselves, and be classified, as black. This traditionally has been the result of rejection by the socio-economically dominant community, basically forcing a child of mixed ancestry into the lower of the two strata of its parents. The paradox is that *racially* you can be one or the other, while *genetically* you can be anything in between. In other words, the heredity of race is not genetic, but social. It correlates with genetics to some extent, but is not genetically transmitted.

Race, in fact, is not even genetically determined. As Madison Grant asserted in *The Passing of the Great Race* in 1916:

> The cross between a white man and an Indian is an Indian; the cross between a white man and a negro is a negro; the cross between a white man and a Hindu is a Hindu; and the cross between any of the three European races and a Jew is a Jew.[17]

Grant was right. But his mistake was in believing that he was making a statement of biological significance. Biologically the statement is nonsense: how could an organism with half its genes derived from one stock be declared a member of another? Grant, rather, was actually making a fairly mundane social observation. The regularity that he expresses here is that where two groups co-exist with significant differentials in power and status, a great deal of weight is placed on relatively small amounts of heredity.

The most graphic example of this comes from societies in which race has taken on extreme social significance—for example, with regard to blacks in America and Jews in Germany. In many American states, "miscegenation" laws were enacted to prohibit intermarriage between blacks and whites. In order to enforce such a law, a definition of the categories was necessary. In many cases (typical were Indiana and Missouri) one was defined as black if one great-grandparent was black.[18] In other words, one-eighth of black ancestry was sufficient to define one as black, while seven-eighths of white ancestry was insufficient to define one as white! The Nuremberg laws established in 1935 not long after the beginning of the short-lived Third Reich, similarly had to define a Jew as one who possessed a relatively small amount of Jewish ancestry.

These laws emphasized not the biological contribution of Jewish or black ancestry to a given person, but its symbolic contribution. In earlier days it was said that one's "blood" was "polluted" or "tainted" by just a small dose of ethnic ancestry. Genetics thus plays a relatively small role in the determination of race: the transmission is mainly through "folk" heredity.

Where racial categories are important in terms of the treatment one receives upon assignment, a great deal of significance is placed on what may be a small genetic contribution. The problem is that the categories are discrete, while the ancestries of people are not. It is the discreteness of these racial categories, in defiance of the biology of the people who are being classified, that makes racial categories fundamentally non-biological. They are social constructs.

Again, let us make clear what we are *not* saying. We are not saying that biological differences among human groups do not exist, nor that racial differences are insignificant. Differences among human groups do indeed exist, but they do not sort the species into a small number of biologically fairly discrete groups. And racial differences are very significant, though not biologically. The question is whether the categories we set up to recognize those differences adequately reflect the biological patterns—or whether they are categories of a different kind. We acknowledge differences among human groups as socially defined and symbolically marked categories, but it is very unclear what underlying biology those categories represent.

The social nature of these categories and their impact can be seen in a recent study by the Federal Centers for Disease Control, which reported significant discrepancies in the statistics on infant mortality. Examining records of babies who were born and died in their first year, researchers found

> I was a student in the Department of Anthropology. At that time, they were teaching that there was absolutely no difference between anybody. They may be teaching that still.
>
> —Kurt Vonnegut, Jr.,
> *Slaughterhouse-Five*

that a small but significant proportion were classified differently on their birth and death certificates. These discrepancies were the result of the system used by the National Center for Health Statistics before 1989, when children were assigned the race of their mother. Before 1989, the process was as follows, according to the story in the *New York Times*: If both parents were white, the baby was white; if one parent was Hawaiian, the baby was Hawaiian; if only one parent was white, the child was assigned the race of its other-than-white parent; if both parents were other than white, the child was assigned its father's race. Apparently this led to the assignment of close to 5 percent of the babies studied as non-white at birth and white at death, resulting in the under-reporting of infant mortality in non-whites.[19]

The most striking demonstration of the social nature of racial categories is to be found by examining a situation where race is important for categorizing people, though with no obvious agenda for exploiting them on that basis: the 1990 United States census forms. Here, with the goal of understanding the racial composition of the United States, respondents were asked in Box 4 to check their race. The categories were White, Black or Negro, American Indian (respondents were asked to print the name of their tribe), Eskimo, Aleut, and Asian or Pacific Islander. Only the last race was broken down into categories, 10 of them: Chinese, Filipino, Hawaiian, Korean, Vietnamese, Japanese, Asian Indian, Samoan, Guamanian, and Other. The last racial category was "Other race (print race)." Box 5 asked a question that wasn't explicitly racial: "Is this person of Spanish/Hispanic origin?" The possible answers were No, Mexican/Chicano, Cuban, and Other (including Argentina, Colombia, Nicaragua, etc.).

These categories obviously are reflecting a single concern: simply breaking the U.S. population down into the groups the government is interested in. Thus, the difference between Irish and Italian ancestry is not considered important, but the difference between Korean and Vietnamese ancestry is. Though no distinction is made among Native Americans, a distinction is made between Eskimos and Aleuts. Hispanic ancestry is separate from race altogether, but is important. No instructions are given for people of mixed ancestry.[20]

These distinctions reflect social categories of concern in contemporary America, where the distinction among national origins is of interest for descendants of Southeast Asians, but not for the descendants of Europeans. The racial distinction between Eskimos and Aleuts is particularly striking, as these are closely related circumpolar populations that even the most ardent anthropological "splitter" would not distinguish from one another as biological races. The concerns here, and the racial categories established to address them, are primarily social.

THE LINNAEAN AND BUFFONIAN FRAMEWORKS

The history of the study of human variation shows that for as long as we have examined ourselves and others, we have been impressed by the differences we encounter. And yet, exactly what to make of those differences has been a source of considerable confusion. It is probably a universal human property to define oneself as a member of a certain group that stands in opposition to other such groups. In a cosmopolitan society, where people whose ancestors came from very different places nevertheless associate with one another, those differences in appearance can be particularly striking. And where wealth and power are unevenly distributed, those differences can certainly reinforce an understandable desire to keep them unevenly distributed.

But what is the underlying biology of group differences? Surely the Linnaean two-dimensional hierarchy of nature dictated that there were elementary clusters of humans to be discerned and named. And these elementary clusters would be the equivalent of zoological subspecies, or races. And yet, it was a surprisingly complicated endeavor when applied to real people in real places.

Were Native Americans a race, more than one race, or a subset of one or more Asian races? Were the dark-skinned peoples of India and Pakistan to be grouped with other dark-skinned peoples (Africans), with the people whom they facially resembled most (Europeans), or with the other peoples of the same continent (Asians)?[21] Were the people of North Africa the same race as the people of sub-Saharan Africa? Within sub-Saharan Africa, were the Khoi-speaking peoples of south Africa (who appeared to have features resembling Asians), the small pygmies of central Africa, the tall and thin Nilotics of east Africa, and the very darkly complexioned peoples of west Africa all in the same race? Ultimately the answers to these questions would have to be arbitrary, for all of these peoples were recognizably different from one another. Can numerically small populations that are morphologically distinct, such as the Ainu of Japan, be considered races, or should that be reserved only for numerically large groups of people—another arbitrary and non-biological judgment?[22] The paradigm under which anthropologists operated demanded that the human species be carved up into a small number of basic units, but the realities of human variation dictated that the breadth of peoples subsumed within each basic unit would render those basic units meaningless.

Our attempts to study human variation have, until the last generation, been formed within the Linnaean paradigm of systematic biology: attempting to establish just how many basic groups there are, and what they are. The assumption under which this paradigm operates is that these few divisions of the human species are basic and profound. The discovery of geographical gradation across human populations, and of variation within human populations, altered the conception of race from a large cluster of human populations to a small set of ideal forms, approximated most closely by the most extreme populations of humans, but present in most humans to greater or lesser degrees.

The criterion of empiricism dictates that scientific work must be grounded in realities, not in abstractions; therefore the scientific analysis of human variation is obliged to study human populations, not abstract human types. Further, the recognitions (1) that race has a large socio-cultural component and (2) that humans have to some extent always migrated and interbred with one another dictates that the appropriate analysis of diversity in the human species lies at the level of the population. We can analyze diversity that exists within population, among populations, or among groups of populations—but higher-order classifications of human populations are largely ephemeral.

Racial analyses have generally proceeded with a flawed conception of human history: that in the obscure past, a few small homogeneous groups of people settled and proliferated, to become large identifiable races, gradation coming as the result of subsequent interbreeding at their margins. There was probably never a time, however, in which the human species existed as a few, small, biologically divergent groups that only later began to interbreed, for this has probably always been occurring.[23]

Under the weight of the synthetic theory of evolution, it became generally appreciated after World War II that human diversity was the result of microevolutionary forces acting on the human gene pool. Human populations, therefore, diverged from one another because of two forces: natural selection (adapting them to different environments) and genetic drift (genetically differentiating them in a non-adaptive manner).

What had been identified as pure races were simply the most extreme human populations. But there is no justification for equating "most extreme" with "pure" or "primeval." The most extreme human populations are simply those which have adapted most successfully to radical environmental circumstances. Thus Frederick Hulse, in the intellectual generation that succeeded Hooton, recast the biological aspects of race in terms of microevolution: to the extent that it had any biological meaning, race became an "evolutionary episode," a transient package of allele combinations and frequencies molded by natural selection and genetic drift.[24] But ultimately, race is not a fundamental biological category at all, as those working within the Linnaean paradigm had assumed. Rather, it appears that human groups can be productively analyzed as populations, but not easily accommodated within a Linnaean framework—as Linnaeus' rival Buffon maintained two centuries ago.

What remained for the anthropology of the last half-century was to focus on the real units of human diversity: populations. And this could be approached using a new set of tools: genetics.

NOTES

1. Hooton (1939a:64).
2. Hooton (1936:513).
3. Castle (1926:147).
4. Ibid., pp. 152–53.
5. Barkan (1992:84).
6. A related phenomenon, obviously, involves prejudging someone on the basis of attributes presumed to be possessed by their sex: i.e., sexism.
7. Dobzhansky (1968), Coon (1968:275). At the time, Dobzhansky was in fact a long-standing member of the American Association of Physical Anthropologists.
8. Livingstone (1962), Brace (1964), Montagu (1964).
9. Vallois (1953); Banton (1987). This kind of thinking has not vanished entirely, interestingly enough, and is represented in the contemporary genetics literature by Nei and Roychoudhury (1993). It appears to be fanciful, however, to imagine a time when there was a small number of discretely different groups of humans who radiated outward from their distant centers of origin and secondarily interbred with other such groups. That is essentially polygenism.
10. Dixon (1923:4). Actually, the work was sufficiently poorly received that Dixon later referred to it only as "the Crime." See Howells (1992).
11. Hooton (1926:78).
12. Ibid., p. 79.
13. Hooton (1946:568ff.).
14. Ripley (1899).
15. Davenport (1917), Herskovits (1927), Davenport and Steggerda (1929), Provine (1973).
16. Terman (1916), Yerkes (1921), Lippman (1922a–e), Klineberg (1935), Chase (1980), Gould (1981).
17. Grant (1916:16).
18. Weinberger (1964).
19. Hahn (1990), Anonymous (1992a, 1992b).
20. On the growing dissatisfaction with the adequacy of formal legal categories to accommodate first-generation or second-generation "multi-racial" people in large numbers, see Shea (1994), Holmes (1994), Wright (1994).
21. Most racial scholars included these people among the "Caucasian" race, but nevertheless continued to define the race as being light-skinned. The implications of including dark peoples among the European racial group were clearly very threatening.
22. Hulse (1955).
23. Boas (1924), Hunt (1959), Shapiro (1961).
24. Hulse (1962).

7

Patterns of Variation in Human Populations

To study biological differences among human groups, cultural or environmental differences must be distinguished from biological ones (e.g., body build, skin color). This is usually hard to do, since both culture and biology affect the expression of any given trait. Sometimes the traits do not even exist, but are simply attributions made by other groups. The human species is polymorphic and polytypic: variation exists within any human group and among human groups. Genetics was expected to reveal the most intimate secrets of human polytypy. Instead, it revealed large amounts of polymorphism.

Under the influence of studies of human history and the synthetic theory of evolution, what were ambiguously identifiable as human races began to be seen as largely ephemeral clusters of alleles. This undercut a fundamental assumption in racial studies, namely, that race was a stable underlying aspect of human variation. But populations were continually mingling (except isolated ones, and by the 20th century there were rather few of those), continually adapting to local circumstances, often fragmenting and coalescing. This new view began to convert racial studies into the studies of human microevolution—which implied that the subject had a much more dynamic component than the previous generation had considered.

A second change occurred in the early 1960s, as biology in general became molecularized. Racial studies had been dominated by studies of the phenotype, the outward manifestations of genes, and had often been guided by a folk view of the hereditary processes; but new methods were being developed for the analysis of genes themselves, or at least their primary products.

THE PHENOTYPE IN RACIAL STUDIES

Studying the phenotype was, to say the least, a confusing way to deal with race. Once it was acknowledged by virtue of simple observation, it not actual fieldwork, that people in any single place looked different from one another, it became necessary to formulate just what kinds of phenotypes were to be used to carve up the human species meaningfully.

Earnest Hooton, in the 1920s, advocated the use of non-adaptive traits in racial analysis: those which offered the bearer neither an advantage nor a disadvantage in the Darwinian contest, but were simply there.[1] The reason was simple: traits favored by natural selection would be expected to arise independently in different populations, thus making them appear superficially to be related. If you wished to establish common ancestry for populations, you would be less likely to be misled by studying "neutral" or non-adaptive traits. But what traits could be considered non-adaptive? How could one reasonably know whether a trait was of selective value or not?

Hooton was well aware that the mere fact that one can make up a story about the utility of a particular trait is not a sufficient indicator that it is an adaptation whose evolution has been guided by natural selection.[2] Thus one could, and presumably should, study human variation in terms of the anatomical minutiae for which physical anthropologists soon became notorious:

> the form, color, and quantity of the hair, and its distribution in tracts; the color of the eyes and the form of the eyelid skin-folds; the form of the nasal cartilages, the form of the lips and of the external ear, the prominence of the chin; the breadth of the head relative to its length; the length of the face; the sutural patterns, the presence or absence of a postglenoid tubercle and pharyngeal fossa or tubercle, prognathism, the form of the incisor teeth; the form of the vertebral border of the scapula, the presence or absence of a supracondyloid process or foramen on the humerus, the length of the forearm relative to the arm; the degree of bowing of the radius and ulna; the length of the leg relative to the thigh.[3]

All of these Hooton considered to be non-adaptive, and thereby useful characters for racial studies. In the second edition of his text *Up from the Ape*, he reversed his stance: "This insistence upon the use of 'non-adaptive' characters in human taxonomy now seems to me to be impractical and erroneous."[4] Regardless of the cause of Hooton's change of mind, his casual reversal on the fundamental question of what kinds of traits actually encode the racial information he sought illustrates the basic problem of racial studies. The criteria did not really matter all that much: Racial categories were real, and obvious, and could be assumed or imposed on data. And yet they defied rigorous definition and diagnosis.

Certainly phenotypes formed the basis of racial studies, indeed of all systematic biology. But which specific phenotypes should be chosen to differentiate races from one another was not entirely clear. There are, broadly speaking, three manners in which phenotypes can differ from one another: First, in their biological development: organisms identical in all respects, but raised at different altitudes, for example, will grow differently. Second, in their non-biological histories: growing animals exhibit considerable plasticity, such that specific stresses (nutritional, for example) or behaviors (such as strapping babies to cradle-boards for transportation) can affect their overall appearance quite significantly. Appropriate differences for racial studies, however, were obliged to come from a third category: constitutional or genetic differences among them.

Often, it is impossible to separate these phenotypic differences from one another. Even worse, sometimes it is difficult to know whether a reported difference is actually real, and not just a fiction, caused less by genes than by something else. Racial odors, for example, have long been noted, but are they real?

In one sense, racial odors are very real, for people smell different. Individuals have their own odors, hence the ability of hounds to track people. Groups of individuals can often be distinguished by their smells, because cultures differ in their habits of personal hygiene, diet, or activity levels—in short, in the many things that go into a phenotype of "aroma." And groups of people invariably find themselves smelling more pleasant than they find other groups of people.

Otto Klineberg, in his 1935 book *Race Differences*, noted that although the 19th-century English found Hindus foul-smelling, the sentiment was in fact reciprocated. In medieval times, a *Foetor Judaeicus* was recognized as emanating from the bodies of Jews. And the Japanese found Europeans notoriously offensive.

But were those smells the secretory products of different group constitutions? Probably not, as people in a cosmopolitan society discover; for even if there is group variation in smell, it is certainly engulfed by cultural behaviors. Hooton was able to dismiss the whole notion anecdotally:

> I once took occasion to ask a brilliant Japanese student of anthropology whether he detected any odor as a distinguishing feature of Whites. He said that he did most decidedly and that he found it very unpleasant. But he went on to say that it particularly assailed his nostrils whenever he entered the Harvard gymnasium. I gave up at once, because I had to admit that his experience coincided with mine. That gymnasium, now happily replaced, was one of the oldest in the country and its entire structure seemed to be permeated by the perspiration of many generations of students. I doubt if the questionnaire method of eliciting information on racial odors will yield satisfactory scientific results.[5]

So ôdors might well differentiate individuals. Group-level odors, however, were probably not constitutional—they were largely subjective, and largely if not wholly "environmental" in origin. They were therefore not useful to the student of biological variation across human groups.

DEVELOPMENTAL PLASTICITY: THE SKULL IN RACIAL STUDIES

Other sorts of differentia were easier to accept than culturally based racial odors. Though odors, the emanations of the body, might not be genetic, it stood to reason that the human body itself was considerably "harder"—differences in bodies were far less likely to be attributable to cultural or behavioral differences. And this was supported by tangible evidence. Before Darwin and Mendel had been born, Buffon speculated on whether raising Africans in Denmark would lead to a decrease in their descendants' pigmentation. By the twentieth century it was a certainty that significant differences in skin color among populations were rooted in stable variants of the genes. The skin, however, was not the organ of greatest interest to students of race. In the words of the German anthropologists, Baur, Fischer, and Lenz: "Brains differ very widely in their functioning according to the degree of civilisation of their possessors, and therefore this organ has always been a centre of anthropological interest."[6]

The fascination with brains and skulls in anthropology is the result of another false syllogism that was widely accepted from the 18th century through the early part of the present century:

The brain contains ideas.
Different peoples have different ideas.
Therefore, the quality of the brain reflects the quality of the ideas the individual possesses.

As Stephen Jay Gould has shown in *The Mismeasure of Man*, the obsession with brains and skulls as a synonym for intelligence took many forms in the scientific study of human diversity. But in the absence of a theory of history or culture (Chapter 3), it did make some degree of sense. The ancient Sumerians or Egyptians, who built civilizations while their contemporaries didn't, obviously had different ideas from those contemporaries, which (it stood to reason) were reflected in different minds, and therefore in different brains and different skulls. It was those brains and skulls that somehow contained the organic basis of civilization, if only they could be analyzed.

In retrospect, this is clearly taking the mechanical philosophy of the Enlightenment too far. Certainly the organic locus of the mind is the brain, as opposed to the pancreas or caecum. But does the brain secrete ideas, as the pancreas secretes insulin, or are they largely received externally from that social and historical stream anthropologists had begun to call culture? As anthropology matured as a science, the latter became recognized as clearly the superior answer. The reason was the appreciation, noted in Chapter 3, of history as a shaper of cultural forms. Studies of acculturation, the contact between cultures, showed that ideas and attitudes are tremendously affected from the outside. Studies of history, both of our own culture and of others, made it clear that people's values and thoughts differed from generation to generation, even when the gene pool of those peoples was held constant.

Before anthropological thought reached this point, however (with no apparent changes in the brain structure of anthropologists), the radical materialism of studying skull differences held a strong appeal. In the 19th century, phrenology was the study of personality by the bumps on the skull. Darwin, indeed, joked later in life about a phrenologist who had predicted that the young Darwin would be an excellent member of the clergy.

Students of the skull across human groups, such as Samuel George Morton and Josiah Nott, without adjusting for age, sex, body size, or nutritional status of their specimens, invariably found the brains of Europeans to be larger than those of other peoples, thus explaining the widespread subjugation of the latter by the former.[7] This measurement was recognizably crude, and the significance of small differences was not clear, though it did reinforce popular ideas about the technological (and therefore intellectual) superiority of Europeans. By the middle of the 19th century, however, a Swedish anatomist named Anders Retzius had developed a method for comparing skulls quantitatively in a more sophisticated manner than simply by gross capacity. This was the cephalic index (Figure 7.1), a determination of the skull's maximum breadth divided by its maximum length, times 100.

The cephalic index was rapidly adopted as a key racial feature. It was quantitative, it was easily measured, it varied consistently across populations, and it involved the skull— ideally suited for a naive anthropological study. By this criterion, the peoples of the world were divided into brachycephalics (those with broad, round heads—a high cranial index) and dolichocephalics (those with narrow, long heads—a low cranial index). Those with cranial indices in the middle, around 80, were mesocephalics. But the system quickly ran into difficulties coping with reality. For example, the Turks were brachycephalic (84), in contrast to the English, but like the Hawaiians and the Siamese. The slightly dolichocephalic English (78) were in the company of the peoples of North Africa and Central Australia. This consequently struck critical observers as an exceedingly artificial way of clustering populations.

Nevertheless it seemed to be a relatively stable marker of populations.

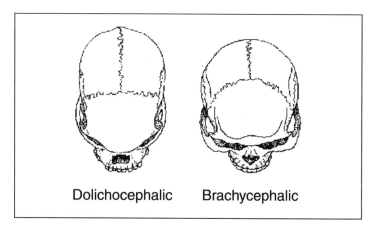

Dolichocephalic   Brachycephalic

Figure 7.1.   Natural variation in human skull shape.

Populations could indeed be characterized by an average skull-shape, and the advent of statistical methods only served to reinforce the usefulness of the measure to distinguish between populations (Figure 7.2). Theories of the evolution of the human species through the repeated migrations and intermarriages of roundheads and longheads circulated.[8] Most of these were founded on a precious little bit of data: that central Europe seems to have more brachycephalic people than northern or southern Europe. Others simply incorporated this bit of information into the old prejudices, making them a bit more scientific.

Some problems were noted, to be sure. For example, one of the advantages of the cephalic index was that it could be measured not simply on the living, but on archaeological skeletal samples as well. And this showed that intact populations, such as the Japanese, had heads that changed significantly over a period of centuries.

> You interest me very much, Mr. Holmes. I had hardly expected so dolichocephalic a skull or such well-marked supra-orbital development. Would you have any objection to my running my finger along your parietal fissure?
>
> —Sir Arthur Conan Doyle,
> *The Hound of the Baskervilles*

There was grudging recognition of a significant cultural component even to skull shape, as various peoples molded the heads of children to make their foreheads high (like the pre-Columbian inhabitants of Peru) or to make their foreheads low (like the pre-Columbian inhabitants of Oregon). Hooton himself found that the extent of cranial deformation varied with stratigraphic level in the sample of prehistoric Americans from Pecos Pueblo. The use of a cradle-board flattened the back of the

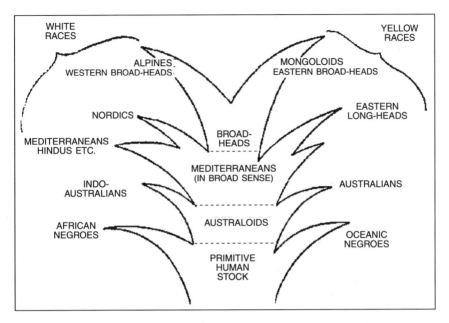

Figure 7.2. Race, rankings, and skulls, in the writing of the distinguished paleontologist A. S. Romer. This figure was reprinted with minor revisions (such as reversing the Australians and Africans) through the 1950s.

head among the aboriginal inhabitants of the American Southwest and of Lebanon.[9]

It was Franz Boas who posed a direct challenge to the use of the cephalic index as a racial measure. If we know that deliberate cultural practices can significantly modify the shape of the skull, he reasoned, might forces other than raw heredity affect the shape of the skull in a more subtle but still direct manner? To study this, Boas designed a classic "nature-nurture" experiment. He saw the rising tide of European immigration into turn-of-the-century America as an opportunity to study the effect of environmental change, while controlling heredity by keeping it constant.

From studying immigrants to America, Boas knew that while some intermarried with other ethnic groups, most had relatives in ethnic communities already in place, and thus remained largely endogamous. By choosing his sample carefully, he could study the effect that coming to America had on the immigrant's body, at the level of the population. His two main target groups were the (brachycephalic) Jews and (dolichocephalic) Sicilians. By measuring the skulls of immigrants and their families already in America, he came to a stunning conclusion: A major component of head shape was determined by the new environment. Jews born in America had longer heads than their incoming relatives and their foreign-born parents, and Sicilians in America had rounder heads than theirs. Further, the extent of difference between immigrant and resident correlated strongly with how long the resident's mother had been living in America before the resident was born. Boas found that several bodily measurements changed with immigration, but the cherished cranial index was the most noteworthy.[10]

Obviously the conclusions had strong political implications, and struck a major blow for the "environment" side, by raising the prospect that social programs might be effective after all, if the form of the body itself is more unstable than had been assumed. But the consequences for the scientific study of race were even more threatening: here was an assumed "hard" biological characteristic of populations shown to be far "softer" than imagined. It is still not known exactly what the environmental stimulus and biological response detected by Boas is, but it is recognizably a consequence of the fact that any growing organism has leeway in its development, and is responsive to many aspects of its environment.[11]

The responses to Boas's work were diverse. The Germans (Baur, Fischer, and Lenz) acknowledged Boas's results but brushed them aside:

These observations in the United States and in German-speaking lands can only be explained on the hypothesis that in particular regions certain environmental influences can directly or indirectly modify the shape of the skull. . . .

On no account, however, must we forget that all such observations and experiments show nothing more than this, that certain elements in the shape of the skull . . . are modified by environmental influences—all the rest remains part of the inalienable hereditary equipment. A delimitation of the two spheres is often impossible.[12]

Hooton downplayed Boas' results by noting that the changes in skull form were on the order of about 2 percent, and it would take a change of at least 5 percent to be significant (despite the fact that the magnitude of change Boas detected amounted to about one-fifth of the range of variation in the entire species). Further, he doubted whether the changes would be as stable as the racial categories were: "It is more likely that an adjustment of the organism to the new environment takes place so that succeeding generations tend to revert to the parental mean or to fluctuate about it."[13]

That very year, however, Hooton's former graduate student Harry Shapiro actually set out to test that hypothesis. Shapiro and another

Hooton student, Frederick Hulse, studied a group of immigrants in a different location, and added a control that Boas lacked. The subjects were immigrants to Hawaii from Japan, compared with Japanese born in Hawaii, and the new controls were the families who remained behind in Japan. They measured literally thousands of people, taking 28 measurements and 41 observations on each, and calculating 21 indices.

Again, Shapiro and Hulse found that the Japanese born in Hawaii differed physically from Japanese immigrants to Hawaii in many measurements and indices, and for the cephalic index, the change was 2.6 points, about six times the standard deviation— indeed, a major change. Given that the children of immigrants did differ physically from their parents, what of the relatives they left behind? Again, Shapiro and Hulse found consistent differences in the same measurements: The immigrants who had been living in Hawaii a short while deviated from their former neighbors back in Japan, the amount of difference being proportional to the length of time since immigration; and those born in Hawaii differed further in the same direction. And there was no evidence of Hooton's prediction of succeeding generations reverting to a racial norm: the next generation diverged further from the ancestral Japanese population than their immigrant parents had.[14]

Clearly this had something to do with the new environment and life: The immigrants lived in a different climate and had different occupations, so again the source of the physical differences was unclear—but their bodies had certainly been modified in ways that undermined basic theories about how to classify and distinguish races. The Japanese immigrants, of course, were not changing so radically as to be confused for Italians—but they were changing, and doing so in the very bodily metrics that were thought to be hereditary and profoundly immutable. Apparently, like deliberate cultural modification of the body, more subtle environmental factors could affect the form of the human body, with more profound consequences the earlier the individual was exposed to them.[15]

GENETICS AND THE HUMAN RACES

By World War II, it was clear that racial studies had reached a crisis. Not only was it apparent that the anthropological work could be egregiously abused to validate the oppression of other peoples, but the very arguments that could be martialed against the *abuse* of anthropological work stood to invalidate its *use* as well. Expert physical anthropologists disagreed over what constituted a race, how many to name, how to identify them, and of what use this information was. What was not doubted was that there existed subgroups of the human species that were genetically distinct from one another. The difficulty was in establishing that genetic distinction. The phenotype, it was becoming clear, was not a reliable indicator of the genotype. In the case of adaptive phenotypes, different populations could achieve the same phenotypic end by different genetic routes; and in the case of non-adaptive traits, the body appeared to be sufficiently plastic that the demonstration of physical similarity or difference did not necessarily imply that the traits were in fact of genetic origin, and thereby "racial."

A way to study "hard" genetic traits was actually discovered early in this century: the ABO blood type system, discovered by Landsteiner in 1900, and which during World War I was found to vary across populations. Its inheritance pattern fit the Mendelian expectations perfectly, which indicated that this was a phenotype resulting from the action of a single gene. By the 1920s it was being applied to the populations of the world by a new kind of student of human diversity, trained in immunological biochemistry rather than in anatomy and physiology.

Races were, of course, regarded as some reflection of the hereditary composition of the human species. The advent of immunological techniques for studying blood groups opened up the possibility of having a direct window on the genes, however small a window it might be. The European cultural mystique about blood and heredity probably aided in the credibility of the serological work—and somewhat uncritically, for it quickly became clear that the claims about the racial study from blood were just insupportable.

In the most bizarre example, a Russian named Manoiloff (Manoilov) reported that a series of simple chemicals added to a sample could reliably distinguish Russian blood from Jewish blood. Directly following this, a disciple reported that Manoiloff's test permitted her to distinguish among the bloods of various Eastern Europeans and Asians. The test turned the blood of Russians reddish, of Jews blue-greenish, of Estonians reddish-brownish, of Poles reddish-greenish, of Koreans reddish-violet, and of Kirghiz bluish-greenish.[16] Though published in *The American Journal of Physical Anthropology*, Hooton found this claim difficult to swallow:

> The results of the Manoiloff test do not inspire confidence. . . . It is inconceivable that all nationalities, which are principally linguistic and political groups, should be racially and physiologically distinct.[17]

By 1929, however, the Manoiloff test was successfully distinguishing sex and sexual preference. Again in *The American Journal of Physical Anthropology*, Manoiloff extolled the work of his colleagues:

> Doctor Livsitz . . . investigated the blood of people imprisoned for sexual crimes. She investigated ten prisoners, one of whom was a sadist, three homosexualists, three suffered from Lesbian love, one was a bisexualist, three had anaesthesia sexualis. In the male sadist and four male homosexualists the Manoilov reaction was feminine; in three women with a unisexual feeling, as well as the one suffering from anaesthesia sexualis, it was untypical masculine. [18]

And it worked just as well on plants, in spite of the biological difficulty posed by extracting blood from them.

A priori, the test made a good deal of sense, as Manoiloff's ideas reveal. "For me," he wrote,

> it is absolutely clear that, by analogy to the presence of hormones characterizing this or that sex, there must be something correspondingly specific of race in the blood of different races of mankind. This specific substance gives the seal of the given race and serves to distinguish one race from another.[19]

Blood is heredity, and heredity is race; therefore, blood is race. What could be simpler? And yet their results were not so much wrong as impossible. Manoiloff was neglecting the fact that "blood" is a *metaphor* for heredity, not heredity itself, and was finding discrete constitutional differences among groups of people he assumed to be constitutionally distinct. Whatever the Manoiloff test was all about, it certainly couldn't do what was claimed. Nothing could: long before there were computers, there was the principle of "garbage in, garbage out."

The analysis of what we now know as the ABO blood group seemed to be on safer ground. Hirschfeld and Hirschfeld (1919) quickly identified three ABO "types": European, Asio-African, and Intermediate, based on the ratio of blood groups A and AB to blood groups B and AB in the populations. Here was a genetic racial classification that conveniently distinguished Europe from the rest of the world. This is, again, historically interesting because we cannot see those patterns now— indeed ABO is taken as paradigmatic of a genetic system in which discrete boundaries among populations or clusters of populations *cannot* be discerned.

The Hirschfelds acknowledged that the inheritance of blood groups "does not correspond with the inheritance of anatomical qualities." In practical terms, it meant that the morphological clusters of human populations were not harmonious with the discernible blood-group clusters: "The Indians [i.e., south Asians], who are looked on as anthropologically nearest to the Europeans, show the greatest difference from them in the blood properties."

But that was not all, for there was prehistory to reconstruct. The presence of three diagnosably different kinds of blood substances (A, B, and O) made it "very difficult to imagine one single place of origin for the human race." Consequently, they proposed an ancestral O human species, subsequently invaded by "two different biochemical races which arose in different places."[20] In other words, heterozygous genotypes were the result exclusively of racial invasions.

By 1925, a student of the blood named Ottenberg could argue under a disclaimer ("My object here is not to draw conclusions concerning an anthropologic question on which I am wholly incompetent to pass") that there were actually six human racial "types" identifiable from the ABO system, at some variance from the groups anthropologists tended to see when they divided up the human species.[21] Laurence Snyder (1926) noted that these investigators had been comparing phenotype ratios, and that the underlying allele frequencies would be a more valuable cross-population comparison. The basic genetics of ABO, as we now recognize it, had only been uncovered the year before: two dominant alleles, A and B, and a recessive allele, O, producing six genotypes and four phenotypes.

Genotypes	AO	AA	BO	BB	AB	OO
Phenotypes	A		B		AB	O

Snyder acknowledged that "the grouping of peoples into 'types' is purely arbitrary,"[22] and nevertheless came up with seven: European, Intermediate, Hunan, Indo-Manchurian, Africo-Malaysian, Pacific-American, and Australian. But he was quick to admit that "because two peoples occur in the same type, it is not implied that they have the same racial history, but only that they contain similar amounts of A and B."[23] Curiously, in spite of a fundamentally different way of analyzing these data, Snyder's sole change from Ottenberg's system was to split "Australian" from "Pacific-American."

The fact is, however, that the groups were not at all distinct. The "European Type," for instance, had percentages of allele A ranging from 19.2 (Iceland) to 34.1 (Sweden), of B ranging from 5.2 (England) to 12.8 (Germany); and of O ranging from 57.8 (Sweden) to 74.6 (Iceland). But three of the 13 populations of the "Intermediate Type" actually fall within this range. Likewise, the "Hunan Type" has allele A ranging from 17.3 to 36.8 percent, allele B from 14.2 to 26.6, and allele O from 42.4 to 66.9. Yet 9 of the 15 "Indo-Manchurian" populations fell within this range.

There was actually no division of these types strictly on the basis of the ABO allele frequencies; what Snyder had produced was a division of the world's populations into large para-continental groups, with the ABO data imposed upon them, and a description of the results. That is why several populations assigned to one "type" actually had ABO frequencies that fell within the ranges of other human "types." Conversely, in some cases, diverse people happened to have too similar a distribution of alleles. This produced a number of inconsistencies: for example, the people of Senegal, Vietnam, and New Guinea ended up together; likewise the people of Poland and China (Figure 7.3).

Hooton, whose primary interest certainly lay in isolating pure racial types, could only muse that

> we can make little or nothing of [blood-group analysis] from the point of view of racial studies. . . . [T]he fact that some of the most physically diverse types of mankind are well nigh indistinguishable from one another in the proportions of the different [alleles], is very discouraging. At present it seems that blood groupings are inherited quite independently of any of the physical features whereby we determine race.[24]

By 1930, Snyder had abandoned the seven race-type system, but still argued "forcibly [for] the value of the blood groups as additional criteria of race-classification." He now had the peoples of the world carved

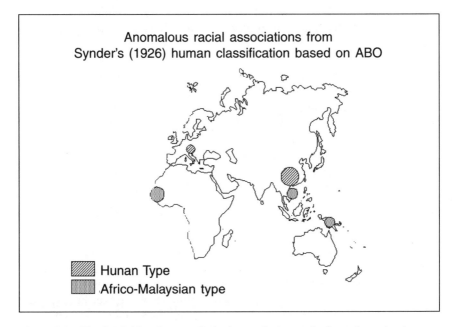

**Anomalous racial associations from
Synder's (1926) human classification based on ABO**

▨ Hunan Type
▦ Africo-Malaysian type

Figure 7.3 The ABO blood group linked populations who by other criteria belonged in different races.

up into 25 (unnamed) clusters, based on different criteria than the phenotypes used by anthropologists, but harmonious to some extent and *dis*harmonious to some extent. His prediction that "in the future no anthropologic study will be complete without a knowledge of the blood group proportions under discussion,"[25] could be considered optimistic, given the specious conclusions that had accompanied its use thus far.

BLOOD GROUP ALLELE FREQUENCIES IN POPULATIONS

Though broad differences can be found in the ABO frequencies across major groups of people (for example, populations of east Asia tend to have between about 15 and 30 percent allele B, while those of Australia have less that 10 percent of that allele), it is obvious that there is extensive overlap. Given a blind test, a series of allele frequencies could not be placed on a particular continent with a high degree of certainty. A large sample of Germans, for example, turns out to have virtually the same allele percentages (A= 29, B = 11, O = 60) as a large sample of New Guineans (A = 29, B = 10, O = 61). A study of Estonians in eastern Europe (A = 26, B = 17, O = 57) finds them nearly identical to Japanese in eastern Asia (A = 28, B = 17, O = 55).

Clearly this single gene is not allowing us to discriminate well among major groups of humans. On the other hand, there is some useful information here. Native Americans have very high proportions of allele O (over 90 percent) and virtually no B; while African pygmies have over 20 percent A, over 20 percent B, and about 50 percent O. Of course, it would be a naive student of human variation, indeed, who would be unable to tell the African pygmy in a group of Native Americans on the basis of phenotype alone! But for the more subtle distinctions among the world's populations, perhaps the addition of another genetic system would allow us to discriminate more clearly among them.

Another blood group system, the MN locus discovered in the 1920s, can be applied. MN has two primary co-dominant alleles, M and N; with it, we find that the Germans have 54 percent M and 46 percent N, while the New Guineans have 6 percent M and 94 percent N. Thus, the two populations that could not be distinguished by their ABO frequencies can indeed be distinguished by their MN frequencies. On the other hand, the Estonians have 60 percent M and 40 percent N, while the Japanese have 54 percent M and 46 percent N.[26] It thus appears that they are not satisfactorily distinct, even using this second gene. Perhaps the addition of a third blood group locus, such as the Rh locus, would allow these two populations to be distinguished from one another.

Perhaps so. Nevertheless, with the addition of more genetic loci, ultimately all populations can be distinguished from all others—since all will have unique constellations of allele frequencies. How, then, does this help to tell what the basic subdivisions of the human species are?

The answer is, of course, that it doesn't; but it took nearly half a century for students of racial genetics to realize it. The earliest students of blood group genetics had immediately inferred from the three alleles in the ABO system the remnants of pure races: a European race of As and an Asian race of Bs, superimposed on a primordial O human race. This naive line of reasoning was roundly rejected by Snyder, whose seven genetical/geographical "types" bore little relationship to genetically pure races.

By the 1940s, however, serologist Alexander Wiener was dividing the world up into three races based on blood frequencies. Somehow, in his hands they managed to sort themselves into a familiar trio: Caucasoid, Negroid, and Mongoloid. Shortly thereafter, he expanded this to six, a classification adopted by William Boyd.[27] But their classification was essentially that of Blumenbach in the 18th century, with a small change: the addition of the Basques of the Pyrenees as a separate race, equivalent to the others—European, African, Asian, Australian, and American. It seems as though the interest in blood groups of human populations added incredibly little to the study of human races. The reason is evident in retrospect: the geneticists weren't *extracting* races from their set of data: they were *imposing* races upon it.

Further, the segregation of the Basques on the basis of their divergent blood groups did not seem to be sublimely wise. It was not as if they had green skin and square heads: they looked like ordinary Europeans, though speaking a strange language and having divergent blood group frequencies. The Basques could hardly be considered a category of living people equivalent to, say, the Africans, and they were certainly not phenotypically distinctive. And the elevation of a single ethnic population to the level of a separate race on the basis of divergent allele frequencies carried other implications. Might the eastern European Jews be a separate race because of their frequency of the Tay-Sachs' disease allele? Or Pennsylvania Amish because of the frequency of their Ellis–van Creveld Syndrome allele? Probably not: the basis of racial analysis was surely the phenotype—to the extent that genotypic studies augmented that, they were welcome. But to base racial distinctions on allele frequencies alone seemed to trivialize the entire endeavor.

As more blood groups and more populations were added, Boyd added a seventh race, "Indo-Dravidian," to accommodate the peoples of south Asia, who were generally united with Europeans by American racial biologists, and divorced from Europeans by English racial biologists. (This demonstrates again the social nature of the categories, given the colonial relationship between England and India.) As still more data came in, Boyd split the groups up further, recognizing the Basques, four other groups of Europeans, a single group of sub-Saharan Africans, a single group of East Asians, the group of Indo-Dravidians, a single group of Native Americans, and four groups of Pacific peoples.[28]

What Boyd was grappling with was the problem of infinite regress: the more genes you look at, the more differences you find among populations. This should have suggested that the proper level of analysis, the unit in human microevolution, was the population, not the race— but somehow it didn't. The scientists themselves were, like others of the generation, prisoners of the consciousness that saw races as the basic biological elements of humanity. And just as it was the task of the student of biological anthropology trained in the analysis of morphology to locate and identify the basic races, so too was it the task of the bio-anthropologist trained in genetics.

But the blood groups do not encode information that permits that question to be answered. Indeed, since the 1960s the work has been cited as *undermining* the concept of race: the pattern one encounters in these data is that of gradual change across space (clinal variation), not of discrete groups separated by clear boundaries (racial variation).[29] This was the same general pattern even Blumenbach had appreciated for morphology, but failed to implement. Yet, like the work of Blumenbach, the genetic work was carried out and interpreted *within* the race concept, and was taken as validating it. The fact that all of the early researchers were able to extract races from the distribution of ABO—in spite of the fact that they aren't there—tells us more about mindset of the researchers than it does about genetic pattern. And the basic pattern was what only Hooton recognized: if you were interested in establishing discrete phenotypic groups of people, the genetic data were pretty much irrelevant.[30]

The depth of the assumptions about races in blood groups can be illustrated by a paradox in this work. If analyzing human races was the goal of studying human variation, and the ideal racial traits were adaptively neutral, then it followed that for the new genetic data to be useful, they would have to be adaptively neutral. And indeed we find the argument being put forth that these genetic data are the ultimate tool for racial analysis precisely for that reason, as Boyd argued through 1950.[31] As the scientific winds shifted, and adaptive traits were now thought to be most useful in postwar racial analysis, Boyd argued still that blood groups were the ultimate tool for racial analysis, precisely because they were *adaptive!*[32] It seems as though whatever kinds of traits were sought, the blood group genes were the best way to study races.

The value of the blood groups in distinguishing races was thus an a priori assumption, regardless of whether they were adaptive or non-adaptive; or whether humans can actually be divided up into a small number of biologically discrete groups. In fact, blood group data were just as ambiguous as traditional morphology in (1) the number of races perceived by investigators; (2) the placement of qualitative boundaries between them; and (3) discrepancies between the groups perceived by these sets of data and by other suites of characteristics.[33]

> Truth often suffers more by the heat of its defenders than from the arguments of its opposers.
>
> —William Penn

The major difference between the genetic and morphological data, however, was that they highlighted an important feature of the heritable variation in the human species, which was also true for morphological data, but often not as obvious. The blood groups were not revealing an allele possessed by all "Mongoloids" or all "Caucasoids" but rather, contrasting proportions of alleles that nearly all populations possessed. Just as any two human groups may differ in average stature, they will nevertheless usually be composed of overlapping ranges of tall people and short people. Likewise, the blood group alleles showed that nearly all populations had A, B, and O; what differed was merely their proportions among populations.

In other words, the majority of biological diversity in the human species was found *within* human groups, not between them.[34] It obviously followed that if one wished to study genetic diversity in the human species, then focusing on between-group differences meant examining just a small part of the scientific problem. The scientific study of human variation had to focus on variations within major human groups, for that is where the bulk of the data would lie. The earlier generations of students, by focusing on the hereditary differences between human populations, had defined for themselves a relatively trivial biological problem.

GENETICS OF THE HUMAN SPECIES

Thus, while genetics was unable to fulfill its promise of resolving the fundamental questions of racial analysis, it revolutionized the field in another way. It ultimately defined that problem out of existence. The study of race would become the study of human microevolution, for race itself was a minor biological issue, involving very little of the diversity in the human gene pool.

This was a conclusion that was easily reconciled to morphological data. The more traits you looked at, the more races you could see, which located race more to the mind of the investigator than to the human gene pool. The proper unit of analysis was the population. Certainly populations carried an evolutionary history, and were related to greater or lesser extents to one another. But identifying the few fundamental biological divisions of the human species was a quest that remained as elusive to genetics as to more traditional methods of investigation. The reason was probably that they weren't there.

The task, then, of the student of genetic diversity in the human species changed in the 1960s and 1970s. The job was no longer to identify and divide the human species into a single-digit number of basic units. Rather, it was to identify the kinds of genetic differences that exist in populations, and to link populations up genealogically to one another. This new goal could be addressed using the fruits of the revolution in molecular genetic technologies of the 1980s.

NOTES

1. Hooton (1926).
2. Hooton (1930b), Bateson (1905), Gould and Lewontin (1979).
3. Hooton (1926:77), Hooton (1931:399).
4. Hooton (1946), p. 452. This shift in Hooton's thought on race parallels a development in evolutionary biology called the "hardening of the synthesis" by Gould. Here, in the postwar era, far more phenomena were taken to be adaptations and less emphasis was given to nonadaptive processes in forming new taxa.
5. Hooton (1931:481). See also Classen (1993).
6. Baur et al. (1931:129).
7. Gould (1981), Michael (1988).
8. Taylor (1921), Dixon (1923), Huntington (1924).
9. Hooton (1930a), Weidenreich (1945), Hughes (1968), Brothwell (1968).
10. Boas (1912).
11. Kaplan (1954), Waddington (1957), Berry (1968), Lasker (1969).
12. Baur et al. (1931:121).
13. Hooton (1931:410).
14. Shapiro (1939).
15. Bogin (1988), Little and Baker (1988).
16. Manoiloff (1927), Poliakowa (1927).
17. Hooton (1931:491).
18. Manoilov (1929:64).
19. Manoiloff (1927:15–16).
20. Hirschfeld and Hirschfeld (1919:677–79).
21. Ottenberg (1925:1395).
22. Snyder (1926:244).
23. Snyder (1926:246).

24. Hooton (1931:490).
25. Snyder (1930:132).
26. Mourant et al. (1976).
27. Wiener (1945, 1948), Boyd (1949, 1950).
28. Boyd (1963).
29. Huxley (1938), Livingstone (1962, 1963).
30. This may explain Hooton's lack of interest in genetics, noted by Joseph Birdsell (1987).
31. Boyd (1950:19, 150).
32. Boyd (1963:1057).
33. The relative merits of phenotypic and serogenetic racial classifications, and the extravagant claims of the latter, were debated in the pages of both the *Southwestern Journal of Anthropology* (Boyd 1947; Rowe 1950), and the *American Journal of Physical Anthropology* (Stewart 1951a,b; Strandskov and Washburn 1951; Birdsell 1952).
34. Lewontin (1972).

8

Human Molecular and Micro-evolutionary Genetics

The principles of genetics are given, from the nucleotide to the gene pool. Hemo-globin genetics is a paradigm for gene structure and function, genome organiza-tion, molecular evolution, and bio-cultural interactions. Though we have learned much about the primary basis of genetic diseases, there are many cultural and ethical questions that remain to be resolved. Genetics is not simply medical; it is cultural as well.

Though the technology and our knowledge of the basic structure of heredity have improved considerably since the days of the eugenics movement, many cultural issues remain unresolved, and loom as large now as they did then. Contemporary genetics allows us to analyze the genetic instructions encoded in DNA, though the implementation of their information in the physiological development of observable phe-notypes is still obscure. Genetics also enables us to determine which human groups, as a result of microevolutionary processes, have higher proportions of certain alleles than other groups. If those alleles cause a specific deformity or disease, the technology exists for identifying it before the affected individual is born. Nevertheless, discussion of the ethical issues surrounding the application of these technologies remains almost as rudimentary as it was in the heyday of eugenics.

GENES AND PROTEINS

The biological molecules that carry out the vast majority of the cell's functions—and ultimately the body's—are proteins, chains composed of virtually any length of any combination of twenty

> **Molecule,** *n.* The ultimate, indivisible unit of matter. It is distinguished from the corpuscle, also the ultimate, indivisible unit of matter, by a closer resemblance to the atom, also the ultimate, indivisible unit of matter.
>
> —Ambrose Bierce

components, the amino acids. Proteins catalyze biochemical reactions, in which case they are considered enzymes; other proteins transport essential molecules, as hemoglobin does for oxygen; others (antibodies) inactivate foreign substances in the bloodstream; still other proteins serve as activation switches for diverse cellular processes. Their ubiquity, significance, and heritability made them reasonable candidates to be considered the actual genetic instructions themselves, but we now know that role to be taken by the more biologically passive DNA.

The heritable variations detectable among the blood groups are protein differences. The difference between a person of blood type A and one of blood type B is due to the proteins (enzymes) that add different terminal sugars to molecules on the surface of red blood cells, or in the case of O, no terminal sugar at all. Ultimately, however, the genes encoding the manufacture of those proteins are composed of DNA—and it is the DNA that is passed down from generation to generation, with the instructions on which proteins for the body to produce. To compare the genetic material and analyze its variations within and across populations involves comparing the products of DNA, or DNA itself.

Although the crucial experiments had been published in 1944, geneticists were slow to accept the revelation that the hereditary instructions in bacteria, and by extension in all organisms, were composed of DNA. Proteins were generally the candidates of choice: they were abundant, diverse, biologically active, and essential to life. Eight years later, experiments on phage (viral microorganisms that parasitize bacteria) once again demonstrated that the instructions appeared to be DNA, not protein. This time, however, a conceptual revolution in biology was launched. The structure of the DNA molecule was reasoned out by Watson and Crick in 1953, and through the 1960s the basic aspects of how DNA functions—the genetic code—were resolved.[1]

The structure of DNA—the famous "double helix"—contains two basic aspects: a structural backbone and an internal sequence of base pairs (Figure 8.1). The pairing of the bases in the center of the helix holds the two strands together, and is highly specific. The sequence of bases (or nucleotides) on one strand determines the sequence on the other strand, so its presence can simply be inferred. It is the precise sequence of bases that literally is the genetic information. Consequently, DNA is often represented schematically as a sequence of letters, each standing for one of the four nucleotides at each position of one strand of the DNA double helix. More detailed information on DNA structure is given in the Appendix.

A gene is, then, a stretch of DNA—a sequence of nucleotide pairs. One strand of DNA contains the informational unit, the gene, while the

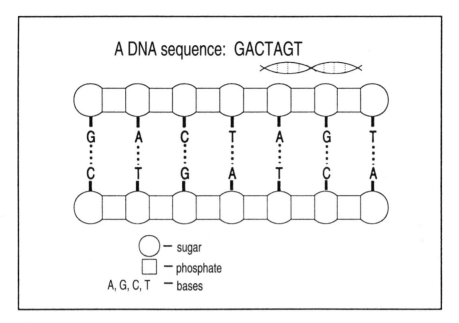

A DNA sequence: GACTAGT

○ — sugar
▢ — phosphate
A, G, C, T — bases

Figure 8.1. The information in a DNA molecule can be summarized as simply the bases on one strand.

other contains its complementary base sequence. The precise definition of a gene is complicated by the diverse nature of genetic information. In general, a gene is regarded as any functional bit of DNA. The most familiar function for DNA is that of coding for proteins, the biologically active molecules in the cell. However, very little of the cell's DNA actually does that.

THE GENOME

The DNA in a typical human gametic cell amounts to approximately 3.2 billion base-pairs (6.4 billion for a cell from the rest of the body). This unit, a single complement of the entire sequence of DNA, is called a genome, and the human genome (and that of most multicellular organisms) contains a far greater quantity of DNA than appears to be strictly necessary. Indeed, relatively little of the DNA appears to be composed of sequences that actually can be considered functional genes; most of the genome is DNA that lies *between* genes. And very little of an average gene is itself "information" (Figure 8.2). Much of a gene's DNA is

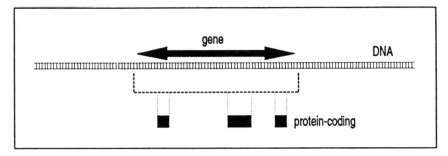

Figure 8.2. A stylized gene: most DNA is not genes, and most of a gene is not
actually genetic "information," translated into protein.

not in fact translated into protein, but is found before, behind, or spliced
out of the actual coding sequence. This has suggested that a goodly portion of the DNA is not used by the cell—most likely genetic "junk," at
least in contrast to our concept of genetic "information."

Though the material spliced out may not itself be used by the cell, the
act of splicing appears to be an important way to regulate the activity
of the gene. Nevertheless, the DNA between genes—what appears to be
a veritable genetic desert with an occasional gene-oasis—is even more
difficult to explain functionally. About 5 percent of the genome seems to
be composed of a 300 bp (base-pair) sequence known as *Alu*, copies of
which are periodically generated and integrated into the genome at
apparently random places. This has been going on for at least 30 million
years, as *Alu* repeats are known from the genomes of monkeys. What is
most interesting about them is that in the human genome there are hundreds of thousands of *Alu* repeats, such that they compose about as
much of the genome as do the protein-coding regions.[2]

*Alu*s constitute a class of "short interspersed elements"—SINES, for
short—and are only one component of the genome. Other parts of the
genome are comprised of localized repeats, simple DNA sequences that
are not interspersed, but arranged in tandem, millions at a time. These
are known as "satellite DNA."[3]

Another reason these genomic components are widely regarded to be
"junk" is that they are very labile. Different species, and different members of the same species, appear to possess widely varying numbers,
locations, and kinds of these repeats. Probably the major conceptual revolution in the last generation in the field of genetics has been the recognition of considerable flux in the genome, belatedly appreciated despite
such suggestions decades ago by corn geneticist Barbara McClintock.[4]

The appreciation of the fluid genome was hindered by the long focus
exclusively on genes, the functional units within the genome. And

genes, it is now widely recognized, are embedded within a complex and dynamic genome, and consequently must be considered as essentially a "special case" of the DNA. The organization, composition, and alteration of genes, in other words, is just a reflection of the general properties of genome organization. Genes are unique in that they happen to be directly responsible for what we observe as phenotypes, but in their milieu they are just more DNA.[5]

The basic mechanisms of change or variation in the genome are still poorly understood, but their effects have begun to be characterized (Figure 8.3). A basic mode of change is the substitution of one base for another in the DNA. Another is the insertion or deletion of one or a few bases. A third is the duplication of a large segment of DNA, creating a tandem repetitive unit—this seems to be the basic way in which gene clusters, spatially proximate groups of genes that are structurally and functionally similar to one another, are built up. A fourth is transposition, the movement of one genomic element to another place, which seems to be a common property of viral sequences in genomes. A fifth is retrotransposition, in which a master DNA sequence makes RNA copies of itself, which are then reverse-transcribed back into DNA, and intercalated into the DNA in diverse places, as appears to be the case with *Alu*. A sixth is caused by mistakes in the copying of short repeated DNA sequences prior to cell division; this "strand slippage" can result

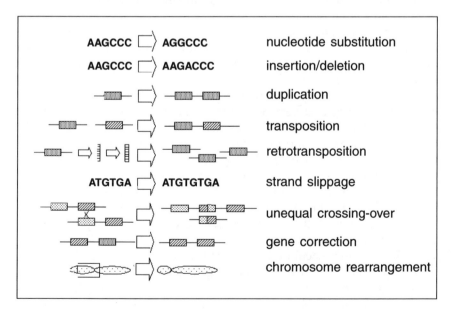

Figure 8.3. Nine general modes of change, or mutation, in the genome.

in a gain or loss in the number of repetitions of the particular sequence. A seventh is unequal crossing-over, in which the normal processes of meiosis produce gains or losses of DNA by virtue of genetic exchanges between genes that are similar in structure, but not true homologs. An eighth is the alteration of a segment of DNA, causing it to conform to a sequence adjacent to it—"gene correction." And a ninth is the large-scale rearrangement of chromosomes, which affects not so much the function of genes, but the way they are packaged

These are the basic ways in which genomes come to differ from one another: the ways in which the genomes of representatives of two species differ, and the manners in which genetic diversity in the human species comes about. Most importantly, however, these processes permit us to see the structure that exists within the human genome, and grasp its origins. This, in turn, allows us to study genetic microevolution.

> Man's yesterday may ne'er be like his morrow;
> Naught may endure but Mutability.
>
> —Percy Bysshe Shelley

HEMOGLOBIN

Like the students of human variation of previous generations, much of our knowledge of molecular genetic variation in the human species comes from the study of blood. Here we use it not so much as a metaphor of heredity, but as a microcosm of heredity. Certainly the best-known genetic system in the higher organisms is hemoglobin, which stands as paradigmatic.

Hemoglobin is composed of two pairs of proteins, alpha or α (141 amino acids long) and beta or β (146 amino acids long). Each of these carries another molecule known as heme, at the center of which is an atom of iron, which is most directly involved in the transport of oxygen throughout the bloodstream. The globin proteins are encoded by two different chromosomal regions, α on chromosome 16 and β on chromosome 11. The composition of hemoglobin is not constant throughout life, however. The 146-amino-acid protein has four distinct varieties: embryonic, fetal, minor adult, and major adult, each of which is produced by a different gene (ϵ—epsilon, $^G\gamma$ and $^A\gamma$— gamma, δ—delta, and β). Likewise, the 141-amino-acid protein has at least two varieties: embryonic (ζ, zeta) and adult (α). There is another gene similar to these, encoding a protein of unknown function (θ, theta). The products of these genes combine to form six different known varieties of hemoglobin, which circulate in the bloodstream at various stages of life (Figure 8.4).[6]

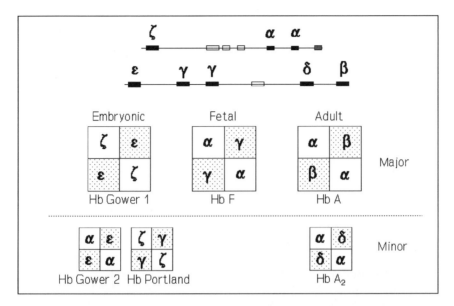

Figure 8.4. Two clusters of hemoglobin genes produce the various hemoglobins circulating in the bloodstream of a normal person at various stages of life. Genes whose products contribute to the known forms of hemoglobin are labeled; other genes are mentioned in the text. For each variety of hemoglobin, the part derived from the β-globin cluster on chromosome 16 is given as stippled.

GENOME STRUCTURE AND EVOLUTION IN THE GLOBIN GENES

The structure of the α-globin gene region shows the effects of genome evolutionary processes at work. Seven genes are present, all of which bear considerable degrees of similarity to one another. Comparing the homologous region in distantly related species shows that there have been repeated duplications of particular genes in different lineages over the eons. The primary evolutionary process inferable, therefore, appears to be a "rubber-stamping," whereby a gene is copied and inserted adjacent to the original.

Three fates exist for a newly duplicated gene. First, in the event that it is advantageous to have more than one gene encoding the same protein, individuals bearing the duplication will be at a reproductive advantage, and natural selection will thereby favor the maintenance of adjacent twin genes. This appears to be responsible for the presence of α2 and α1, which encode precisely the same protein. Second, having another gene may not be advantageous, and so the duplicate may accumulate mutations over the generations, with no ill effects either—as long

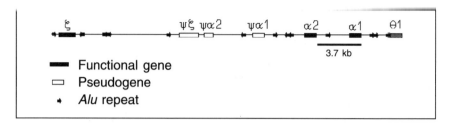

■ **Functional gene**
▢ **Pseudogene**
✦ *Alu* repeat

Figure 8.5. Fine structure of the α-globin gene cluster, showing regions of serial homology: genes and *Alu* repeats. The entire region spans about 30 kilobases.

as the first gene remains structurally intact. Ultimately, fortuitous mutations may result in the production of a protein that has somewhat different specificities from the original, and may then take on a stage-specific or tissue-specific role. This appears to be the origin of the differences among ζ, α, and θ. And finally, the second gene may simply accumulate mutations that render it inoperative, in which case it becomes just intergenic DNA bearing a resemblance to neighboring gene sequences. In this case, the sequence is called a *pseudogene*. We can see in the α-globin region a single ζ pseudogene and two α pseudogenes.

Distributed throughout the α-globins are several *Alu* repeats, which have been interposed within this gene region (Figure 8.5). The double-stranded nature of DNA permits them to be integrated in two opposite orientations, one in which the sequence read on just one strand ends in AAAAAAA, and the other in which it begins with TTTTTTT. In the latter case, the *Alu* sequence is read properly from the other strand of DNA, and consequently is considered to bear an opposing orientation.

A few regions of short tandem repeats can also be found in this region. The recently silenced pseudogene of the embryonic zeta bears a mutation that truncates the protein it produces. This mutation is not present in the chimpanzee, and is indeed polymorphic in humans. Within the first internal non-coding region (intron), zeta has 12 tandem repeats of the sequence ACAGTGGGGAGGGG, while its nearly identical pseudogene has 39. In individuals who have two functional zetas (due to a process of homogenization of adjacent sequences called "gene correction"), both have 16 copies.[7]

THE COMPARISON OF GENETIC REGIONS

The most obvious way to compare genetic regions directly is by establishing their nucleotide sequence, using methods that have become commonplace in the last few years. Now one can study the genotype directly; indeed, since non-genic DNA comprises most of the genome, this implies the overarching irrelevance of these data to the study of phenotypes. The main advantage here is the fact that a specific genetic region is being studied in great detail. The main disadvantage is that the procedure of sequencing DNA is labor intensive, and one often relies on very small sample sizes (this is particularly a problem where comparisons of DNA between species are being made).

The use of the polymerase chain reaction (PCR) permits the amplification (that is, the production of usable quantities) of very specific short sequences of DNA that can be directly sequenced. This allows greater numbers of subjects to be studied without greatly extending the work required. It also permits minute quantities of DNA to be analyzed in forensic studies.

As an alternative to sequencing, one can use a battery of DNA-cutting enzymes known as restriction enzymes, which cleave DNA very precisely at defined sequences of nucleotides. The enzyme HindIII, for example, cuts DNA everywhere it encounters the sequence AAGCTT (Figure 8.6).[8] Given the size of the human genome, the enzyme will make many cuts, but these will be very precise. If a single base differs in a recognition sequence (say, AAGCTT mutated to AAGCTC), the enzyme will bypass that site, and rather than produce a particular DNA fragment of the expected length, will yield a substantially larger DNA fragment. The ability of geneticists to detect these differences in cutting sites provides a way of surveying large genomic regions for variants in a small number of nucleotides, and surveying large numbers of people. Additionally, this provides a manner of examining the particular genomic "surroundings" of a defined region of DNA.

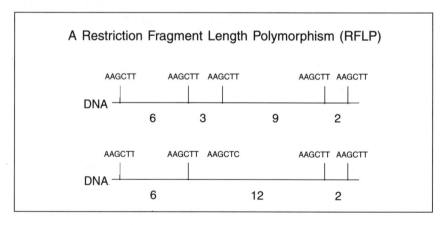

Figure 8.6.   RFLPs can reveal nucleotide differences among individuals, if a single base change alters an enzymes's recognition site. Numbers represent lengths of DNA in kilobases.

Sometimes a length variant is detected that is not due to a point mutation at a restriction site, but to a quantitative change in the amount of DNA present. This may be the insertion of a brand-new *Alu* repeat.[9] In some cases, however, it may occur in a region of redundancy, such as noted in the ζ-globin region. In this case, the sequence variant may be a change in the number of repetitive elements present. This would be a variable number of tandem repeats, or VNTR.

HEMOGLOBIN VARIATION IN THE HUMAN SPECIES

The fundamental difference between genotype and phenotype is highly visible in the hemoglobin genes. Hundreds of variant genes have been discovered within the human gene pool. Though each differs minimally from the others, by virtue of a nucleotide substitution, its effects upon the body range from better-than-normal, to no effect at all, to debilitating disease. Genetic variation in non-genic DNA will be generally unexpressed; and changes in non-coding regions of genes are usually also phenotypically silent. Changes in the DNA coding sequences range from those which do not change an amino acid in the hemoglobin protein (again, phenotypically silent), to those which encode a variant amino acid with properties similar to the original (generally benign), to those in which the structural and functional integrity of the protein is compromised by virtue of the amino acid substitution. In homozygous form this can be potentially lethal.

The paradigmatic genetic disease is sickle-cell anemia, in which a nucleotide substitution near the beginning of the beta-globin gene causes a variant protein to be produced. This minuscule difference gives the protein variant electrochemical properties, which in turn affect the interaction of the hemoglobin molecules in red blood cells.

The cells, generally smooth and rounded, develop sharp irregular edges when packed with sickle-hemoglobin, and clog the capillaries. The resulting problems of circulation damage the spleen, heart, and brain principally, and obviously many other bodily processes are impaired as well.

The sickle-cell allele is most common in people from west Africa, among whom it may reach a frequency of over 25 percent. The reason for the high frequency of such an obviously harmful allele is the fact that the allele is benign when the bearer possesses only one copy; that is, when the bearer is a heterozygote. Indeed, it is better than benign: it mitigates the effects of a severe endemic blood disease, malaria, such that heterozygotes living in malarial areas have a better chance of surviving and reproducing than "normal" individuals. This is known as

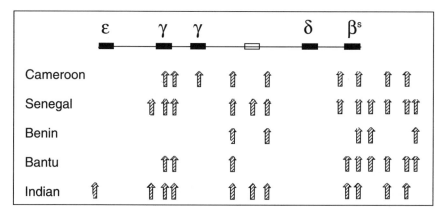

Figure 8.7. A map of restriction sites shows that the sickle cell mutation has arisen several times in several genomic surroundings or *haplotypes*. Each is named for a geographical area where it is common.

"balancing selection," in which nature has struck a balance between the beneficial effects of both alleles together in single dose, and harmful effects in double dose of either allele separately.[10]

Studies of restriction fragment length polymorphisms indicate that the sickle cell allele resides in five different genomic environments, which suggests that it has arisen and reached significant frequency five different times in different populations.[11] In Figure 8.7, the sickle-cell allele is designated β^s, and the arrows denote sites cleaved by specific restriction enzymes. The five regional types are all particular combinations of nucleotide sequences revealed by the property of being cut, or not, by the enzyme. Two of the cut-sites are common to all five sickle-cell allele types; the rest create unique patterns within which the sickle mutation has arisen.

Where malaria does not exist, sickle cell loses its heterozygote advantage and becomes simply a debilitating genetic disease to homozygotes. In America about 1 in 13 African-Americans is a sickle-cell heterozygote. Though this population is empirically at greatest risk, sickle-cell is known from other populations as well, notably those of the Mediterranean.

THALASSEMIA

Just as sickle-cell rose in frequency because of the advantage it conferred to heterozygotes, so too have other blood diseases that appear to confer similar immunity to malaria. Other alleles encoding different structural variants of the beta-globin proteins are known as hemoglobins C and E (sickle-cell is widely called "hemoglobin S" or "HbS"). These are all qualitative hemoglobinopathies: the body produces a structurally abnormal protein.

Another class of genetic diseases of hemoglobin is known as *thalassemia*. Here, the basic phenotype involves hemoglobin that is structurally intact, but diminished in quantity. In the beta-thalassemias, generally a mutation has occurred to a DNA region regulating the transcription of the gene to its mRNA. As a result, the gene is transcribed inefficiently. The mRNA is reduced in quantity, and the beta-globin protein is as well.[12]

In alpha-thalassemia, a deficiency in the amount of alpha-globin is generally due to the absence (via deletion) of an entire alpha-globin gene. By virtue of their location in a cluster of structurally similar DNA regions, an inaccurate "crossing-over" during meiosis can produce a chromosome with only one or no functional α-globin genes rather than two. The heterogeneous phenotype of the disease is now known to be the result of the diverse genotypes (ranging from four functioning α-globin genes to none). And the heterozygous form of the thalassemia genotype appears to serve the same function among the inhabitants of southeast Asia that the sickle-cell allele serves among the inhabitants of west Africa—a genetic adaptation to malaria.[13]

GENETIC SCREENING

Gene pools may be characterized by high frequencies of usually rare disease alleles for one of two reasons: Either they confer an adaptive advantage for survival and reproduction in certain situations in single dose; or they proliferate by virtue of chance factors, operating through the variable of population size, and overcoming the allele's harmful effects by virtue of the "founder effect" (Chapter 2). In either case, the widespread prevalence of a harmful allele raises a medical issue. For example, in the case of sickle-cell anemia, given that malaria is not a major health threat in contemporary America, might we not spare a family tragedy and correct the 1/4 of the offspring of the union of 1 in 150 African American couples whom we would expect to be at risk for sickle-cell anemia?

This is, of course, not a problem confined to African-Americans. With 1 in 30 Jews of eastern European ancestry a carrier for Tay-Sachs' disease, and 1 in 25 northern Europeans a carrier of cystic fibrosis, the application of technology to childbirth and family planning has an impact on all people in developed countries, and impends over less economically developed countries as well.

Genetic screening has a eugenic goal: the reduction or elimination of genetic disease. But the goal here is different from the goal of eugenics in the 1920s. Then, the goal was to assist the "race" by ridding it of nebulously-defined undesirables. Now the goal is to assist the *family* in bearing healthy children, who will not be seriously debilitated or die from the combination of alleles they inherit; and to give the family the option of not giving birth to a genetically disabled baby.

This is not, however, strictly a medical issue. If the goal is to target members of a specific ethnic group for screening, one needs to decide whom to include, and how to establish that they are in fact the right subjects. One needs to establish who pays—whether the state guarantees the right of a couple to have children as free of genetic disease as is technologically possible, or whether that is only for those with the means to pay for it. Also, abortion has to be an option, for if a diseased fetus cannot be aborted, then there is not much sense in a genetic screening program. Further, the goals of the program and theory of Mendelian genetics must be set forth clearly and intelligibly. One would not wish to leave a prospective couple with the idea that they are to be sterilized if the test comes out positive, for example.[14]

Indeed, sometimes the goals are not very clear even to the scientists: in the days when sickle cell carrier status was detectable, but there was no test available for a fetus *in utero*, a prominent scientist seriously suggested that heterozygotes be branded on the forehead, so that all prospective mates would know the individual's genotype. To that scientist the purpose of screening was apparently to uncover heterozygotes; to most others, in fact, the goal is to uncover homozygotes—uncovering the heterozygotes is simply a means to that end.[15]

The example raises another interesting issue, however: access to the information. Presumably your genotype is private information. But since there have been numerous scandals involving illegal access to information on credit histories, does it seem reasonable to expect that information on genotypes will be less accessible? And the possibility is also always present of prejudice on the basis of genotype, a situation that has already arisen with respect to insurance companies.[16] Often, since the individuals screened are members of targeted ethnic groups that have been traditionally subjected to discrimination, the very attempt to reduce the burden on a family can result paradoxically in discrimination.

Other fairly common genetic defects are congenital, though not passed on from parent to child, such as Down's syndrome, and variations on the sex chromosome complement [XXY (Klinefelter's), XO (Turner's), and XYY]. Screening for carriers is useless, for there aren't any: affected individuals are born to genetically normal persons. Down's poses a higher risk to the fetuses of women over 40, but a woman over 40 who wishes to have a child may feel she will not have another chance to get pregnant, and thus may be disinclined to lose the fetus. Here, the rights of the mother may come into conflict with those of the fetus and society. A widely publicized situation involved a newscaster with an autosomal dominant allele causing a deformity of the digits, and a 50/50 chance of passing it on. She nevertheless chose to have a baby with the condition—to a chorus of public disapproval.[17]

There are thus many cultural values that go into the genetic screening program. These include the idea that conscious control over the health of the baby is desirable; that certain genetic conditions are readily diagnosable and should be prevented; and that certain reciprocal obligations exist between the citizen (either the couple or just the mother) and the state.

MODERN EUGENICS

Eugenics is still controversial, as the narrower medical goal still intersects with a number of significant social issues, such as abortion, control of women's bodies, and access to genetic information. Do people have an obligation to have healthy babies? If so, to whom? Does a baby have the right to be brought into the world without deformities—such that a parent who deliberately bears a child known to have such deformities can be considered abusive? Does the state have the responsibility to care for preventably genetically handicapped babies? What constitutes a handicap? Mental impairment? To what extent? Physical deformity without mental impairment? Physical handicap without deformity? If prospective parents know that their fetus will require considerable medical support from the state to thrive, does the state (or the taxpayer) have the right to insist that the mother try again for a healthy baby? Does the state have the responsibility to guarantee the health of babies by regulating prenatal behavior of mothers, which is a far greater cause of congenital problems than parental genotypes are?[18]

The eugenics movement is no longer with us, though the word remains in use, with a somewhat narrower range of applications. Whereas in the 1920s the focus of the eugenics movement was on the improvement of the race, and subsumed the sterilization of entire classes and ethnic groups on the basis of a casually inferred genetic inferiority, the locus of contemporary eugenics is the family, and its aim is the identification and prevention of more objectively identifiable genetic disease.

The old view of eugenics had to be discarded in the face of mid-century advances in the study of genetics. The assumption that favorable traits cluster in certain groups of people and are entirely absent in others had to be abandoned to the recognition of diversity, which must be considered "favorable." The assumption that bringing the human species closer to a uniform genotype would be desirable had to be abandoned to the recognition that genetic diversity is abundant and apparently necessary in natural populations. The possibility that heterozygotes may be generally fitter than homozygotes undermines the eugenic doctrine, as the reproduction of heterozygotes continually replenishes the supply of less fit homozygotes. Further, as it is now fairly clear that human groups have adapted genetically to local environments, of which endemic malaria is simply one example, genetic diversity in the species is too valuable a commodity to dismiss lightly.

And, of course, our appreciation for the limits of the control that genes have over our bodies has heightened. While we all appreciate the genetic basis of many types of medical pathology, many of whose etiologies have now been established in nucleotide changes, it has proven far more elusive to establish the genetic basis for traits falling in the wide range of "normal" physical and behavioral variation. Though genetic links to alcoholism and schizophrenia, for example, are often cited and may well exist, their relation to phenotypes is inconsistent. Thus one could at best identify individuals with a significant chance of developing the phenotype to the specific genetic background, and miss many others who develop the phenotype for other reasons. The course of action to take in such a probabilistic situation is unclear.

HEREDITARIANISM

What remains with us, almost unchanged since the time of the eugenics movement, is the idea that one can reasonably posit a gene for virtually any human condition that can be expressed in a noun. This hereditarianism is older than eugenics, and older than genetics, though it can always be framed in the language of contemporary science. In modern genetics, hereditarianism takes root in the clinical nature of the data—in the relationship between pathology and normality. Are there genes for, say, aggression? Or self-mutilation? Who can say for certain? But the conceptual problem here lies in the simple preposition, "for." There are certainly genes that *affect* those qualities. Lesch-Nyhan syndrome, for example, is caused by a mutation on the X-chromosome. Affected children have a terrible and tragic compulsion to bite—to bite their lips and fingertips off—and they do so if not permanently restrained. If that is the disease, then what is the gene for?

Alas, we don't know—at least we can't tell from the disease. The gene makes an enzyme involved in the metabolism of purines (i.e., bases in DNA). That much we do know. Then physiology intervenes; and out the other end of the black box of organismal development comes a dramatic, bizarre phenotype.[19]

Lesch-Nyhan serves as an instructive cautionary tale about genetics. Knowing the pathological phenotype of a mutant may tell you little or nothing about the function of the normal allele. And knowing about the genetic pathology may tell you little about the phenotype in the general population. This genetic syndrome has aggressive self-mutilation as a phenotype, but is there any sense to the statement that there is a gene controlling self-mutilation? Certainly adults who harm themselves[20] are not at all suffering from Lesch-Nyhan, so we can learn virtually nothing about the general behavior from the study of this genetic pathology.

Indeed, almost all of our knowledge of contemporary human genetics comes from studying diseases. While this is obviously very valuable information, it is important to acknowledge the kind of information that does *not* come readily from these studies. Often, for example, we do not learn what a gene *does*, only what the pathological phenotype is. It is tempting to speculate that the opposite of the pathological phenotype is the gene's normal role. This would, however, imply that the function of the normal allele for Lesch-Nyhan Syndrome is to *prevent* you from biting off your fingertips. A strange job for a gene; a strange conception for the nature of human biology, where the body's normal state is to bite off one's fingertips unless restrained by this gene.

Imagine, by analogy, trying to discern the function of an automobile carburetor by randomly smashing it with a hammer and observing the effects. You might notice the color of the exhaust changing; but it would not be valid to deduce that the function of the carburetor is to regulate the color of the exhaust fumes, or that the normal color of exhaust is black, unless acted upon by a carburetor.

It is far easier to understand how a system can be broken down than to understand how it works. Since most of our information on human genes involves pathologies, it should not be surprising to note that the vast majority of genes with known phenotypes are diseases (Figure 8.8). The other large category of genes, which overlaps this, are the biochemical minutiae like hemoglobins, antigens, and enzymes, whose phenotypes are often understandable only at the level of the biochemicals themselves. We know nothing of the genes for height, body build, nose shape, hair color—in short, of the genes for the normal range of human phenotypic variation. As far as our understanding of human phenotypes goes, we have only their pathological breakdown products.

Consequently, the reports of advances in our understanding of molecular genetics can be quite misleading. Genes are often named by the disease their alleles cause. One hears about the gene "for" Huntington's chorea, or cystic fibrosis, or tumor suppression—how they have been

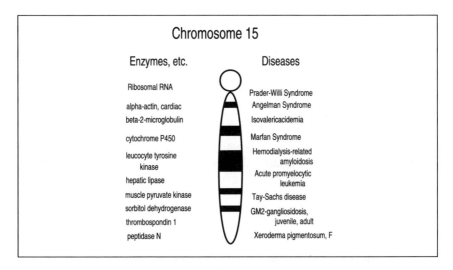

Figure 8.8. Major genes mapped to human chromosome 15, from O'Brien (1993).

located, isolated, and analyzed.[21] But this carries an odd message about human biology. Can the normal state of the human body be to make tumors, such that a gene is required to suppress it? In what sense is there really a gene "for" cystic fibrosis? Is the body so morbidly built that we are loaded down with genes simply there to destroy us? Hardly. The genes are simply named by virtue of the pathological phenotype resulting from their mutations, but their "normal" function is understood poorly, if at all; and the genetic basis of phenotypes that vary widely and normally in human populations is virtually unknown.

Retardation is the most common phenotype associated with genetic disease in humans. Affected individuals with phenylketonuria (PKU), for example, have diminished intelligence.[22] Is this a gene for intelligence, then? Or is it simply that human intelligence is physiologically precarious, and can be damaged very easily by a wide variety of genetic (and also, obviously, environmental) factors? Clearly the latter; for the effects of PKU on intelligence are side effects, or pleiotropies. Obviously the genes affect intelligence. The paradox, though, is that although we know of genes that can radically diminish intelligence, we know nothing of the normal range of variation for that trait in the gene pool. Knowing about PKU tells us nothing about the math whiz, the philosopher, or the dropout.

Nevertheless, a recent study claimed to have found a gene for aggression.[23] A rare biochemical variant for an enzyme called monoamine oxidase A, segregating in a Dutch family, was associated with strong tendencies to physical violence. A gene for aggression? Certainly a gene producing aggression when its function is compromised. But it tells us nothing about the large number of violent crimes in society, the vast majority of which are by committed criminals who lack that allele. Understanding the rare pathology connects only in a very tenuous way to an understanding of violence generally, of cultural variation in violence, or to understanding why normal people sometimes do violent things.

It is unlikely that molecular genetics is going to lead us back into the era of eugenics—into sterilizing the underclass— for the cultural milieus of the eras are quite different. Nevertheless, there are many ancillary issues requiring thought and some degree of historical reflection. For example, the editor of *Nature* writes:

> The truth is . . . that geneticists themselves are likely to be the first to recognize the dangers of interfering with the natural flow of genes within a population before the social implications are understood. Indeed, only geneticists can recognize the dangers.[24]

"What is truth?" asked Pontius Pilate. Certainly the truth expressed in the above passage is not borne out well by history. Practitioners of genetics were among the *last* to appreciate the implications of what they were teaching and taking for granted in the 1920s. Why would they be first now? Although the ideas and the technologies have changed, the past is often the key to the present. Here the issue is whether the science of human genetics is objective and value-free, or whether there are always cultural assumptions camouflaged and invisible to practitioners, but only recognizable from a distance—either historical or intellectual distance. In the latter case, the scientists themselves would be the least likely to be able to distinguish their science from the cultural values being promoted *in* their science—which is indeed what history shows us. Actually, it is simply a truism of anthropology: to understand thoughts and deeds comprehensively requires a frame of reference outside the specific system of ideas producing them.

NOTES

1. Hotchkiss (1979), Judson (1979), McCarty (1985), Wallace (1992).
2. Schmid and Jelinek (1982), Schmid and Shen (1985), Shen et al. (1991).
3. Britten and Kohne (1968), Willard (1991).
4. McClintock (1950, 1956), Temin and Engels (1984), Cavalier-Smith (1985), Weiner et al. (1986).
5. Hunkapiller et al. (1982), Kao (1985), Marks (1992b).

6. Honig and Adams (1986).
7. Ohno (1970), Marks (1989).
8. A key feature of the DNA sequences recognized by these enzymes is that they are palindromes. The sequence AAGCTT is given from 5″ to 3″ (see Appendix); the other DNA strand is not only complementary in sequence (TTCGAA), but also opposite in its orientation (3″ to 5″). When read in the 5″ to 3″ direction, the sequence of both strands is AAGCTT.
9. Trabuchet et al. (1987), Wallace et al. (1991), Perna et al (1992).
10. Neel (1949), Pauling et al. (1949), Ingram (1957), Livingstone (1958), Honig and Adams (1986).
11. Kan and Dozy (1980), Lapoumeroulie et al. (1992). Alternatively, it is conceivable that the mutation arose just once, and the five haplotypes represent the products of extensive mutations and crossing-overs.
12. Serjeant (1992). Many diverse kinds of mutations are known to cause beta-thalassemia, including some that diminish the efficiency of translation, rather than transcription.
13. Higgs and Weatherall (1983), Collins and Weissman (1984), Bank (1985), Flint et al. (1986), Honig and Adams (1986).
14. Rapp (1988, 1990), Motulsky (1989), Duster (1990), Marfatia et al. (1990).
15. Pauling (1968), Duster (1990), p. 46.
16. Billings et al. (1992), McEwen and Reilly (1992), Natowicz et al. (1992), Allen and Ostrer (1993), Alper and Natowicz (1993), McEwen et al. (1993).
17. Seligmann and Foote (1991), Hubbard and Wald (1993), Rennie (1994).
18. On bioethics and the genetic counseling issues raised by the Human Genome Project, see Capron (1990), Hubbard (1990), Suzuki and Knudtson (1990), Miringoff (1991), Cowan (1992), and Parker (1994), Garver and Garver (1994).
19. Stout and Caskey (1985), Stout and Caskey (1988), Sculley et al. (1992).
20. See, for example, Russell (1993).
21. Ponder (1990), Stanbridge (1990), Solomon (1990), Collins (1992), Roberts (1992a), Travis (1993), Kinzler and Vogelstein (1993). A related example can be seen in a paper that follows a recent trend of using declarative sentences as titles: "Hox11 Controls the Genesis of the Spleen" (Roberts et al., 1994). The mouse homeobox gene examined is involved in the formation of the spleen, for when the gene is absent, so is the spleen. But that is "control" only in a very narrow and unconventional sense of the term, meaning a necessary but not sufficient condition for the spleen's development.
22. Sriver et al. (1988), Eisensmith and Woo (1991).
23. Brunner et al. (1993), Morell (1993).
24. Maddox (1993).

9

Human Diversity in the Light of Modern Genetics

The genetic work that was expected to provide the validation for dividing humans into races instead undermined it. Genetic variation is to the largest extent polymorphism, and not polytypism. The fundamental units of the human species are populations, not races. Nevertheless, populations differ from one another in the frequencies of the same alleles they carry, and this can be used to group human populations by genetic similarity. Like morphological differences across the human species, these groupings are generally obvious when extremes are contrasted, but otherwise there is little in the way of reliable biological history to be inferred by genetics.

Though the eugenics movement is no longer with us, we still maintain an interest in genetic medicine and in classifying people. Though they have changed over the course of the 20th century, both of these interests remain important: genetic medicine as a manner of improving the lives of individuals and families; classification as a cultural means of self-identification. Both of these are potentially useful: the former, as long as a concrete genotype and pathological phenotype are well-defined, and the links between them are very clear; and the latter, as long as the groups with which one identifies are not presumed to reflect discrete biological categories. Unfortunately, we retain a strong cultural tendency to "see" three discretely and fundamentally different groups of people in America: blacks ("African-Americans"), whites ("Caucasians"), yellows ("Asian-Americans"), and when pressed, reds ("Native Americans").

Yet the categorization is easily undermined when another group, Hispanics or Chicanos, is added. For now the criterion of inclusion is not the purported continent of ancestral origin, but the language spoken by one's ancestors. One can have significant ancestry from Europe, Africa, and/or the New World and be Hispanic, for it is a "racial" category that transcends "race."[1]

Even geneticists, as products of their culture, occasionally still write of the "three races" as if they were genetically marked from one another. It is neither surprising nor scandalous, but merely convenient, though it reflects little of biological significance to the student of the human species.

DIFFERENCES AMONG THE "THREE RACES"

Populations differ from one another, and those differences are geographically patterned. People tend to be more similar to those people who live nearby, and more different from those who live far away. Fairly obviously, then, people plucked from very disparate locations will be found to vary from one another substantially. People plucked from neighboring areas will also be found to vary, but more subtly. Microevolutionary processes have been at work: natural selection differentiating populations and adapting them to local conditions; genetic drift differentiating populations in random, non-adaptive ways; and gene flow homogenizing populations. With people differing subtly from their close neighbors, it becomes difficult to imagine how the human species could be effectively and objectively carved up into a small number of biological units, or races.

Since the aboriginal populations of the world do differ from each other in distinctive ways, most obviously in pigmentation and facial features, it is often possible to allocate individuals to one of the major groups of immigrants to America (Figure 9.1). These immigrants, significantly, are derived from geographically localized regions of the Old World. The populations of the world are heterogeneous and intergrading, but if one compares people from very different places, one finds them, unsurprisingly, looking very different.

Contemporary forensic

Figure 9.1. Major zones of migration from the Old World to America, resulting in the appearance of three discrete races.

anthropologists are often asked to identify skeletal remains as to race. Here, knowing the ways in which people vary around the world can assist us in establishing the "race" of an unknown skeleton. Obviously we use the word "race" guardedly: we are simply saying that if we divide the ancestors of living Americans into three categories, we can make a better-than-random guess about which of them an unknown skeleton falls into. This is not to suggest that there are three clear biological categories of people: only, rather, that three populations from widely different parts of the world can be distinguished from one another.

Some of the distinguishing characteristics of the skull involve the wide and projecting cheekbones of "Asians"; nasal projection of "Europeans"; and wide distance between the eye orbits of "Africans." These characteristics overlap between groups, and are quite variable within each of the three groups; but armed with a list of such average differences, anthropologists can fairly reliably allocate skulls into those three categories, or more. Table 9.1 lists criteria provided by a forensic anthropologist to assist in allocating specific skulls to one of five groups.[2]

The purpose of such an exercise is to assist law enforcement officials by providing them with additional information about a murder victim. None of the traits is perfectly diagnostic; these are average differences, and do not imply fundamental divisions of the human species into a small number of basic homogeneous types. Other criteria are also diagnostically useful, such as the shape of the femur, which tends to be more straight in "Africans" and more bowed in "Asians." Again, however, it is crucial to appreciate that this does not mean that there are three discrete biological categories of people. It means simply that, *given* three categories, skeletal remains can reliably be assigned to one or another of them. This is a consequence of two facts: human populations differ from one another, and Americans are derived generally from large groups of immigrants from geographically distinct areas.

Imagine a child given a set of blocks of different sizes, and told to sort them into "large" and "small." Not only would the child successfully allocate them, but the most extreme blocks would be invariably allocated into the same category by different children, while there might be a bit of discordance over the allocation of some of the blocks in the middle. The fact that the blocks can be sorted into the categories given, however, does not imply that there are two kinds of blocks in the universe, large and small—and that the child has uncovered a transcendent pattern in the sizing of blocks. It means simply that if categories are given, they can be imposed upon the blocks.

Table 9.1. Criteria for Allocating Skulls to Different Human Groups (after Gill 1986)

Traits	Mongoloid	American Indian	Caucasoid	Polynesian	Negroid
Cranial form	broad	medium-broad	medium	highly variable	long
Sagittal outline	high and globular	medium-low, sloping frontal	high, rounded	medium	highly variable, post-bregmatic depression
Nasal form	medium	medium	narrow	medium	broad
Nasal bone size	small	medium-large	large	medium	medium-small
Nasal profile	concave	concavo-convex	straight	concave/concavo-convex	straight/ concave
Nasal spine	medium	medium, tilted	prominent, straight	highly variable	reduced
Nasal sill	medium	medium	sharp	dull/absent	dull/absent
Incisor form	shovelled	shovelled	blade	blade	blade
Facial prognathism	moderate	moderate	low	moderate	high
Alveolar prognathism	moderate	moderate	low	moderate	high
Malar form	projecting	projecting	not projecting	projecting	not projecting
Palate form	parabolic/elliptic	elliptic	parabolic	parabolic/elliptic	hyperbolic
Orbital form	round	rhomboid	rhomboid	rhomboid	round
Mandible	robust	robust	medium	robust, rocker form	gracile, oblique gonial angle
Chin projection	moderate	moderate	prominent	moderate	small
Chin form	median	median	bilateral	median	median

THE SOCIAL NATURE OF GEOGRAPHICAL CATEGORIES

The "three races," then, merely designate three major migrations into the United States: from (West) Africa; (Western) Europe; and (East) Asia. The indigenous peoples of Eurasia, however, blend gradually into one another, and the indigenous peoples of Africa blend into those of the Near East, and are themselves physically very diverse. Indeed a contemporary social phenomenon popularly called "Afrocentrism" involves the appropriation of "Africa" as a homogeneous racial and cultural entity. "Was Cleopatra Black?" asked the cover of *Newsweek* in 1989.

There are no natural boundaries separating the people of Europe from those of Asia, and the one that appears to separate Africa from Europe— the Mediterranean—is far more permeable than it appears, and has been successfully navigated for thousands of years. The most formidable natural boundary actually subdivides Africa: the Sahara desert. People from north of the Sahara look far more like southern Europeans than like equatorial Africans. Thus the category "Africans" is itself a cultural construct, artificially lumping together highly diverse peoples.

On the eastern side of the African continent, the Nile has long connected equatorial Africa with Egypt; consequently the Near East has long been a biologically highly cosmopolitan area. Was Cleopatra black? It is hard to say, but contemporary images depict her as looking rather like contemporary of inhabitants of the Near East do (Figure 9.2). As a member of an intermarrying Macedonian dynasty, she probably more closely resembled a modern-day Egyptian than a modern-day resident of, say, Ghana or Denmark.

Once again, however, we confront here the overlay of cultural values upon ostensibly racial or biological categories. The category "African," as in "African-Americans," really means Central-West Africans, the people whose ancestors were brought to the New World as slaves. One would not ordinarily consider the descendant of an Arabic-speaking, Muslim Egyptian as falling into this category. So just asking the question "Was Cleopatra Black?" involves substituting the entire geographical continent Africa for the region of Central-West Africa, and

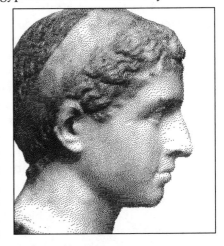

Figure 9.2. Portrait of Cleopatra VII (69–30 B.C.), after Antiken museum SMPK 1976. 10, Berlin.

thereby implying that Africa is composed of a relatively homogeneous population. But regardless of whether or not Cleopatra was black—that is to say, whether she resembled a modern-day African-American—the question clearly means a great deal more to modern Americans than it did to Caesar or Antony.

Likewise, our category "Asians" refers really to immigrants who arrived in America principally from east, and generally southeast, Asia. Though the people in the indigenous lands of Asia blend gradually into one another, "Asian-Americans" are drawn from a more localized geographical area. When the intervening populations are omitted or ignored, descendants of southeast Asians, west Europeans, and west Africans certainly provide a stark morphological contrast to one another.

And the category of "White," "Caucasian," or "European" is no longer subdivided. In 1939, Carleton Coon saw and described 10 races within the white race. Others more commonly saw three. While we no longer classify individuals or populations as "Nordic," "Alpine," or "Mediterranean," it is important to appreciate that lumping them into a single category is itself a cultural artifice that has two main consequences. First, it acknowledges the superficiality of the human differences that exist within and across the European continent; and second, it sets up an easy contrast to "African" and "Asian."

The fact is that just as the categories of "European," "African," and "Asian" obscure subdivisions that blend into one another, or gradients in aboriginal biological diversity within each category, so too do they mask the fact that these three major categories also blend into one another across the aboriginal geography.

"Racial" categories thus divide by nomenclature people who cannot be easily divided from one another biologically in the Old World, except in the extremes. In America, these categories are useful for classifying groups of immigrants, but they do not represent fundamental biological divisions in our species—they represent, rather, only biological patterns perturbed by social and historical forces. Those biological patterns are principally geographical gradients, upon which we have tended to impose discrete cultural boundaries. People from the same part of the world tend to look more like each other than they look like people from a very different part of the world; but there are no natural borders around them.

An obvious demonstration lies with the Jews, who are united *by definition* culturally, rather than biologically—and who were long considered to be a "racial" issue. If race is a strictly biological category, and Jews are a strictly cultural category, then there should be no sense at all in a phrase like "the Jewish race." And yet both Jews and non-Jews alike can identify people who "look Jewish." Is this a contradiction? Not really: it simply reflects the fact that a significant proportion of Jews (particularly in America) have ancestry from southeastern Europe, and consequently tend to look more like one another than like people from Norway or Pakistan.

And yet, many Jews do not "look Jewish," and many non-Jews do. This reflects the other side of the coin: that after generations of gene flow and religious conversion, the Jews of Yemen look like Yemenis, and the Jews of Spain like Spaniards—and that most people from southeastern Europe are not Jewish. This conclusion is borne out as well by genetics, which finds populations of Jews from one aboriginal region almost invariably to be very similar to populations of non-Jews from the same region; and often more similar than to Jewish populations from elsewhere.[3]

Anthropological genetics, which was developed in order to validate racial categories—to find a hard hereditary basis by which to divide the human species—was never able to do so. Despite the fact that there is a hereditary basis for phenotypic differences—alleles by which, say, blond people differ from brunet people—these have not yet been found. Indeed, as we noted in the last chapter, exceedingly little is known about the genetic basis of "normal" phenotypic variation in the human species. And is blond/brunet a racial difference? Linnaeus defined Europeans as blond (Chapter 3), but of course most Europeans are not. And some darkly complexioned Australians are.

What we do know about genetic variation in our species is comprised of two main categories, as we noted in Chapter 8. The first is pathologies. Since most contemporary genetic work in humans is motivated by medical goals, it follows that most of what we know about genetic variation involves ways in which the human body fails to function properly, leading to cystic fibrosis, phenylketonuria (PKU), Tay-Sachs disease, or any other of a host of genetic pathologies. The second category is that of biochemical minutiae, such as the blood cell antigens, of which the ABO blood group is most prominent. Many of these are involved in cellular recognition processes—notably the highly polymorphic histocompatibility or HLA loci—while others are simply variant forms of enzymes whose overall efficiency is neither helped much nor hurt much by the biochemical difference.

The most extreme of these biochemical minutiae is the restriction fragment length polymorphism, or RFLP, which we described in Chapter 8. RFLPs are DNA segments defined structurally, not functionally, with reference to the length of the DNA segment produced when a specific restriction enzyme is applied to cut the DNA. In a given region of the genome, applying the enzyme EcoRI (which cuts DNA at the sequence GAATTC) may result in the DNA of interest being cut at GAATTC sequences 2000 nucleotides apart. If some individuals have a different nucleotide in place of one of the six in this particular EcoRI recognition site, the enzyme will not cut the DNA there, and will instead cut the region of interest at the "next" site. This will make the length of the restriction fragment produced by the application of this specific enzyme in this specific region appear somewhat longer. Differences among individuals that are detectable in this way are restriction fragment length polymorphisms, or RFLPs.

We can imagine a DNA segment defined by enzyme cut sites 2000 base pairs apart. A single nucleotide change in the recognition sequence will cause the enzyme to fail to cut the DNA there. In an experiment designed to detect the *nearest* restriction site to the original (an experiment known as a Southern blot), the DNA segment may now appear to be 3000 nucleotides in length. Though the cut site 3000 bases away may be present in the original subjects, this experiment detects only the *nearest* site, and therefore ordinarily ignores the third site, which becomes the nearest in the other subjects. Thus, this is an RFLP with two alleles, a 2000-nucleotide-long variant, and a 3000-nucleotide-long variant. These can be distinguished by their migration in an electric field, in which short DNA fragments move farther and faster than longer ones. Looking at the DNA of 6 people (Figure 9.3, left to right), the first and

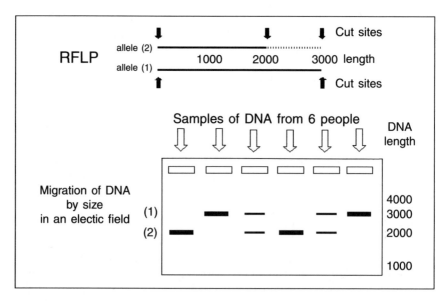

Figure 9.3. Detection of an RFLP, segregating as a pair of Mendelian alleles in a population.

fourth are homozygous for the 2000 allele, the second and sixth are homozygous for the 3000 allele, and the third and fifth are heterozygous.

By virtue of this minimal genetic change, a single nucleotide substitution, it is possible to ascertain differences among people, and among populations. By using many enzymes and many regions, a fairly detailed picture of the pattern of genetic diversity in the species can be established. These changes will have nothing to do with the observable phenotypic differences manifested by people or populations, since they are usually not being detected in gene-coding regions, but they will be useful estimators of the patterns of *genotypic* differentiation undergone by our species. These data, at the most fundamental genetic levels—the presence of one nucleotide versus another—reinforce what was established in the 1960s from cruder genetic comparisons based on proteins: genetic polymorphism in the human species is far greater than polytypism. In other words, most genetic variations are found in most populations, though in varying proportions.[4] The study of human genetic variation, then, is principally the study of diversity within populations; to focus on genetic differences *between* populations is to define a very narrow and biologically trivial question.

PATTERNS OF GENETIC DIFFERENTIATION

At present there is no way known for genetics to establish ethnic groupings based on genotypes from a large series of individuals. The reason is that there are two major categories of polymorphic variants that are found in the human species. The first consists of polymorphic alleles that exist in all, or nearly all populations, such as the blood groups. This is the most common form of genetic variant in the species—ubiquitous, but in different amounts from population to population. The second major category of polymorphic variant is known as a "private polymorphism"—an allelic variant found only within a restricted population, or a restricted part of the species, though not characterizing all, or even most, of its members.

The "Diego antigen," for example, is a genetically-determined blood cell antigen. The allele for it (Di+) is found in Asian and Native American populations exclusively—so in that sense it is specific to those populations. Nevertheless, the proportion of people possessing the Di+ allele varies only from less than 10 percent (in some East Asian and North American groups) to over 40 percent (in some South American populations). However, the allele is entirely absent from other South American populations. And, of course, most people even in the populations of which it is characteristic, are Di–. Presumably the allele arose in Asia and was carried over by some of the early immigrants to America. As their populations expanded and fragmented, genetic drift elevated the frequency in some populations and reduced it in others; in most, small amounts of gene flow maintained the polymorphism at a low level.[5]

Perhaps the most extreme example of such a private polymorphism is the "Duffy" blood group, for which three alleles exist: Fy^a, Fy^b, and Fy^4.[6] The first two alleles are the only ones present among many populations of the world, across Europe, Asia, and the Americas. They vary somewhat in frequency: a sample of Thais revealed the Fy^a allele at 83 percent, and the Fy^b allele at 17 percent; a sample of Germans had the Fy^a allele at 42 percent, and the Fy^b allele at 58 percent; and a sample of aboriginal Australians had the Fy^a allele at 100 percent. The third allele, however, is most common in Africans, particularly in people from central and west Africa, where it is virtually the only allele present. Nevertheless, the Fy^4 allele is present in southwest Asia in appreciable frequencies, and the other two alleles are common in East Africa. In general, about 20 percent of the "Duffy" alleles in Americans classified as black are Fy^a and Fy^b, and as those alleles are co-dominant, generally over 1/3 of black Americans tested have *phenotypes* of this blood group antigen not found in central or west Africa.

Far more common, however, are the ubiquitous polymorphisms. Even in DNA segments that presumably only originated once and have no conceivable effect on the phenotype, we find the DNA to be polymorphic all over the world. A graphic example involves an *Alu* repeat (Chapter 8) that has been integrated into a specific spot in the gene for "tissue plasminogen activator." Consequently, some copies of this gene are 300 base-pairs longer than others, and they can be studied as a simple length variant with two alleles. In a coarsely divided sample of people, the *Alu*-present allele had a frequency of 66 percent in the genotypes of "Asians," 63 percent in "Caucasians," and 42 percent in "African blacks." It was subsequently found at 10–16 percent in the genotypes of aboriginal Australians, 12–20 percent in Papua New Guinea, 29–58 percent in Indonesia, and 58 percent in Japan.[7] The fact that the allele was generated by an insertion makes it likely that this originated as a single mutation, not a recurrent one; and the fact that it is everywhere makes it difficult to consider these gene pools as very distinct from one another.

Ethnic groups are therefore not genetically marked as races in the manner that an earlier generation of European and American scholars believed. The historical biology of human populations is the result of the forces of microevolution, and these have not produced obvious genetic clusters corresponding to what we would identify as a race. There can thus be no genetic test to perform in order to determine whether or not one is "Caucasian," "Alpine," or "Hopi." The reason is simple: populations are constantly in genetic contact with one another. The only populations in which alleles rare elsewhere have come to characterize all members of a specific group are small, isolated island populations. In other groups of people there is a constant and dynamic flow of genetic material, adding diversity to the gene pool, sometimes only a little each generation, but guaranteeing heterogeneity in any gene pool. Additionally, widespread polymorphism may be due in part to ancient variation carried over in descendant populations, which reinforces the deep roots of genetic heterogeneity in human gene pools.

Presumably when groups are distinguished on the basis of phenotypes, there is usually an underlying genetic basis—in the case of skin color, for example. Genetically, however, this proves exceedingly vexing, for three reasons: (1) We do not know how many genes contribute to the phenotype and how many segregating alleles there are for each gene. (2) Dark skin color is found aboriginally among many peoples who do not appear to be closely related, for example, Pakistanis, Australians, and Central Africans—and we do not know whether the genes and alleles underlying it are identical in these cases. (3) The skin color of "black" peoples and of "white" peoples is actually highly variable, both aboriginally and because of interbreeding. This makes it impossible even to conceive of a genetic test for assigning people to groups even on the basis of this classical, relatively clear-cut phenotype.

The general assignment of humans to groups on the basis of their hereditary makeup could be done with some degree of built-in arbitrariness, and that would be a trivial matter were it not for a complication in the social universe. Different groups often receive different treatment and have different opportunities. This makes group membership far more significant than the genetic underpinning of it would justify, for (as we have already seen) heredity is quantitative, while group membership is an all-or-nothing affair.

If we could envision a society in which people were judged on the basis of their own accomplishments, rather than on the group to which they are assigned or with which they identify themselves, such a society could maintain races, and yet not be racist. That is presumably the kind of society we strive for in the cosmopolitan, industrialized 21st century.

Certainly we gravitate to people to whom we perceive ourselves to be similar; this may be one of the most fundamental human drives. And the kinds of bonds we form that way are symbolic: whether we gravitate to other Mormons, or other Baltimore Orioles fans, or other citizens of Irish ancestry, we form associations based on the perception of shared feelings. The problem involves formalizing these associations, such that people who do not share some specific quality of interest are thereby barred, or deprived of basic rights. And the appropriation of *genetics* as a basis for group membership, or more significantly, for group exclusion, is a pernicious misapplication of human genetics. It is what we mean by "racism."

The resolution of the problem of racism is not to deny group differences, which obviously exist; nor to deny the human urge to associate with like-minded people, which is undeniably strong; but to ensure that the diverse groups of people in contemporary society are given equal access to resources and opportunities. In other words, to assure that individuals are judged as individuals, and not as group members. The opportunity for self-improvement is vital to a free and cosmopolitan society, and the possibility to take advantage of it must be independent of group considerations.

This highlights, we may note, one of the most fundamental errors made by the eugenic social theorists. They maintained that group heredity overrode individual genotypes, phenotypes, or potentialities. Consequently, the apparent fact that the (genetically inferior) lower classes were out-reproducing the (genetically superior) upper classes caused them concern.

When we acknowledge, however, the equivalence (though not identity) of gene pools across social classes, and (as we will see in Chapter 11) the demographic trends by which entry into the middle class and higher education generally lead to a reduction in family size, the "swamping" problem perceived by the eugenicists evaporates. There will always be smart and talented people to run the country. Intelligence and talent take many forms. We simply have to cultivate them from the lower classes, and give them the opportunity to express their talents. Many of the current generation's talented people, of course, are derived from lower classes of earlier generations—the very ones that the eugenicists feared and loathed. It is hard enough to run a bureaucracy, much less a civilization, with *exclusively* the "best and the brightest"—imagine what a needless burden is placed on society by *failing* to cultivate the abilities of large segments of the populace!

The ultimate and paradoxical end of the racist eugenics program would have been to create a society of the second-best and second-brightest. They would have utterly failed to detect the talented people, by virtue of focusing on group biology (often, of course, pseudo-biology), rather than on the biological gifts and potentials of individuals.

MITOCHONDRIAL EVE

Certainly the most celebrated study of human genetic diversity in recent years has been the "mitochondrial Eve" study by Rebecca Cann, Mark Stoneking, and Allan Wilson, in 1987. Mitochondrial DNA (or mtDNA) is the little bit of hereditary material that exists and functions outside the nucleus of the cell, in a cytoplasmic organelle called the mitochondrion, whose function is to generate metabolic energy to fuel biochemical processes.

Studying the presence of restriction enzyme cut sites distributed around the 16,500 base-pairs of the circular mtDNA molecule, they calculated the minimal number of genetic changes that had occurred to the mtDNAs of about 150 people from all over the world. They fed the information into a computer to estimate a phylogenetic history for these DNA sequences. The results are quite clear, that it is difficult to locate any obvious patterns. People from the same local population generally have similar mtDNAs, but there are often other people from widely different populations scattered among them. Certainly no clear higher-order clusters, which might indicate genetically homogeneous "races," are evident.[8]

The other controversial and highly publicized claims cited for the work are based on two trivial inferences. First, that the mtDNA sequences are descended from a single person, and second, that the person was female. Though both are true, they mean little. All mutations, after all, originate in single individuals; the analysis of the diversity in the DNA sequences in effect involved extrapolating backwards to the single original sequence, which necessarily existed within a single original person (Figure 9.4). Of course that person was *not* the only human alive at the time; merely the only person at the time whose mitochondrial DNA has been passed down to the present generation.

Further, mtDNA is not transmitted in a Mendelian fashion, in which a child is equally closely related to mother and father, with sperm and egg contributing the same quantity of chromosomes. Mitochondrial heredity, by contrast, occurs strictly through the egg. In other

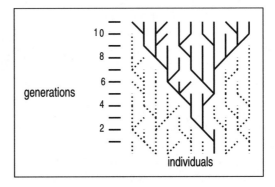

Figure 9.4. With a constant population size, and each person having 0, 1, or 2 daughters, a single mtDNA type soon characterizes the entire population.

words, a child is mitochondrially a clone of mother, and unrelated to father. Certainly, then, since we have extrapolated backwards to find the single individual who transmitted all modern mtDNA down to the present generation, and men do not transmit mtDNA to their children, it follows (since it was passed on) that the founding sequence existed in a female body. Further, the tidy "tree" derived from mtDNA inherited clonally and maternally will have little in common with the actual genetic relationships of the organisms, extensively mixed up by virtue of the Mendelian, biparental processes.

The students of mtDNA also calculated a rate by which mtDNA appeared to change within human populations, and estimated that the founding sequence existed 200,000 years ago, a date roughly congruent with the origin of anatomically modern humans. The significance of this congruence, however, is not immediately clear, since there is no necessary relationship between the origin of *Homo sapiens sapiens* and the origin of variation specifically in mtDNA. The connection between them is rather more subtle. The origin of variation in some DNA sequences, such as the histocompatibility or HLA genes, appears to be far older, pre-dating the divergence of humans and apes. The origin of variation in other DNA sequences, such as the hemoglobin genes, appears to be far more recent, and probably tied to the widespread adoption of irrigation (Figure 9.5)

For the student of human diversity, however, the mtDNA work represents the best genetic survey of the human species to date. It managed to generate strong support for two inferences about genetic variation in the human species drawn from previous studies, and also supported by subsequent studies. First, as noted, one does not encounter racial clusters within the data. At best, there are clusters of people from the same population, but no clear patterns beyond that. And second, if the continental origin of the bearer is imposed on the distribution of the genetic variants, one finds far more genetic diversity in (sub-Saharan) Africa than in other comparable geographic regions (Figure 9.6).[9] This in turn implies that genetic diversity has been accumulating longer in sub-Saharan Africa than elsewhere.

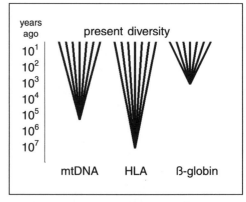

Figure 9.5. The origin and spread of diversity in a specific gene is variable, the result of specific evolutionary forces operating on it.

The latter inference is the genetic basis for the "Out of Africa" hypothesis, which derives the diverse population of modern humans from a founding population in sub-Saharan Africa, which then expanded outward to colonize the entire Old World. The regional differentiation of modern humans would thus be very recent geologically, subsequent to the emergence of modern humans. Its alternative would hold that regional differentiation of modern humans is far more ancient, inherited from regional differentiation in the fossil species that preceded our own (i.e., *Homo erectus*). This was, it may be recalled, the thesis of Carleton Coon in his controversial *The Origin of Races* (1962).

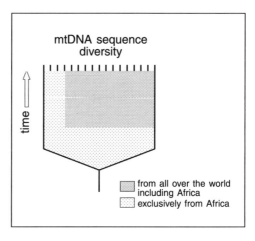

Figure 9.6. Africa appears to have the most mtDNA diversity and the most extreme variants, suggesting that mtDNA has been there longer than elsewhere.

While genetic data cannot resolve these issues absolutely, it does appear that differences among human groups genetically are less extensive than would be predicted under the assumption that this diversity has preceded the emergence of our species. It seems to be fairly recent in origin. Indeed, the amount of genetic diversity encountered within the human species is generally far less than encountered within our closest relatives, chimpanzees and gorillas, which have presumably been accumulating genetic diversity for roughly the same amount of time (Figure 9.7).[10] Why would the human species be so depauperate in genetic diversity, in comparison with our closest relatives? Presumably the explanation lies in the demographic history of our species. Something happened that cut back on the diversity we now find, so that what we encounter now is relatively recent in origin. The most likely

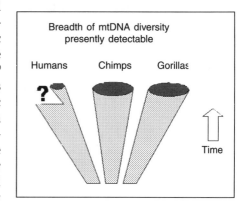

Figure 9.7. The relative amount of genetic diversity in humans suggests a recent cutback.

candidate process is a "founder effect," in which a relatively small group of people ultimately became the progenitors of a large group of descendants. These founder effects are often associated with speciation events—the descendants become a new species.

If the human species really is depauperate in genetic variation relative to chimpanzees and gorillas, and if this really is the result of the founder effect, then it may well be the case that the present level of genetic diversity in our species dates to the emergence of our group, anatomically modern *Homo sapiens*, about 200,000 years ago.

PATTERNS OF GENETIC DIVERSITY

One interesting mtDNA polymorphism demonstrating the general pattern of human genetic diversity is reflected in a length difference of 9 bases. The longer sequence is the only allele found throughout Europe and Africa. In Asia, however, an appreciable number of people have another allele—the shorter sequence. Nevertheless, those who have this "Asian allele" are distinctly a minority: 18 percent of mainland east Asians surveyed, and 16 percent of Japanese. In some Polynesian and Micronesian island populations, the "Asian allele" can assume a very high frequency, even approaching 100 percent of the people on some islands; while in Melanesian islands, such as Papua New Guinea, or in the Americas, the allele can achieve a frequency higher than many Asian populations, though the people are quite distinct from them.[11]

Though the shorter DNA sequence is regarded as an "Asian allele," most Asians do not in fact have it; and an individual who has it may have ancestry from Asia, America, or the Pacific. And there is no telling how remote the Asian, American, or Pacific ancestry of an individual with the shorter sequence might be. Thus, it is of no use as a "racial marker"; rather, it is a private polymorphism with a restricted distribution, as many seem to be.[12]

The other large category of polymorphic genetic variation, as we have seen, involves alleles present in virtually all populations at some appreciable frequency. This may be due to the polymorphism being present in the ancestors of the present-day species, and being passively inherited by contemporary populations; to the long-term admixture of populations over many generations; or, less likely, to recurrent mutation, coupled with selection. It is virtually impossible to distinguish among these explanations in any particular case.

Mitochondrial DNA also shows this broad pattern of ubiquitous distribution. Using the presence or absence of restriction enzyme cut sites as alleles (or "morphs"), we find that the enzyme AvaII detects an allele called morph 1. Morph 1 is found in 89 percent of the Maya of central America, 55 percent of Senegalese in west Africa, and 12 percent of south African San. The allele known as morph 3 is there in 11 percent of the Maya, 1 percent of the Senegalese, and 59 percent of the San. Private polymorphisms make up the rest of the African sample.[13] Widely divergent populations thus turn out to be remarkably qualitatively similar in their detectable genetic composition.

Genetics thus gives us a deep insight into the hereditary differences distributed within the human species: so far as they are detectable, they are between person and person, and far less between group and

> The human species, according to the best theory I can form of it, is composed of two distinct races, the men who borrow, and the men who lend.
>
> —Charles Lamb

group. Further, the establishment of large or basic biological races is not in the least bit clarified by the introduction of these data. There is no evidence for a primordial division of the human species into a small number of genetic clusters that are different from one another. The fact is, we do not know how many basic groups of people there are, and it is very likely that there is *no* small number of groups into which a significant proportion of the biological diversity in the human species collapses.

THE GENETICS OF INDIVIDUALITY

Genetically we are all different from one another, with the trivial exception of identical twins. Further, genetic diversity appears to be of considerable importance to survival and reproduction. The theoretical geneticist Sewall Wright developed much of modern population genetics based on the idea that any population has an optimum frequency of different alleles at the same locus segregating within its gene pool.[14] This would in turn imply a great deal of heterozygosity on the part of the organisms composing the population. The more empirically-minded geneticist Theodosius Dobzhansky found extensive heterozygosity among natural populations of fruitflies, and indeed found heterozygotes to be better survivors and reproducers than homozygotes.[15]

Perhaps the strongest evidence for the necessity of genetic variation has been found in cheetahs: these once-widespread predators have considerable difficulty in both surviving (being very susceptible to infectious diseases) and reproducing (having high proportions of defective sperm). Various methods of measuring genetic diversity showed that cheetahs are virtual clones of one another, having almost no genetic variation, presumably the result of repeated population crashes. In the histocompatibility genes, which determine whether a skin graft will "take" or not, the high levels of genetic diversity in the human species and in other mammals ensure that (without immuno-suppressive drugs) skin cannot be grafted from one individual to another—except possibly between close family members. In the case of cheetahs, however, there is no barrier to the grafting of skin among individuals: they appear to be genetically identical to one another even for these genes.[16]

It appears, then, that genetic diversity in a population is necessary for the optimal state of the gene pool. It ensures heterozygosity in individual organisms, which (by virtue of cryptic physiological processes) confers benefits over homozygotes.

Once again, therefore, we see that a basic assumption of the eugenics movement was flawed. The eugenicists assumed there was a single best homozygous "type" toward which humans could and should be bred. Presumably the best contemporary example of such a product is the cheetah, seriously endangered on account of its very homozygosity.

That mode of thought, which holds there to be an ideal form (or genotype) against which all others are degenerates, inferiors, or trivialities, harks back to Plato. Its modern replacement is that there is a broad range of normality and a multitude of genetic potentials that come in all combinations in all people. In one area of contemporary science, however, that former view still holds sway—in the clinically-oriented area of molecular genetics.

THE HUMAN GENOME PROJECT

The advent of DNA sequencing techniques, which permitted the precise determination of genotypes by virtue of reading their sequence of nucleotides in a linear sequence, revolutionized biology in the 1980s. One consequence of this revolution was to mobilize resources for a massive project to determine the nucleotide sequence of the entire human genome—that is, the 3.2 billion bases composing the genetic information in a single cell. This would not only tell us what we "really" are, but cure cancer and lead us to economic recovery as well. Molecular geneticists lobbied Congress, and enthusiastic biochemists and science journalists sang the praises publicly of ultimate self-realization through DNA sequence.[17]

The success of DNA sequencing lay in the diagnosis of individuals at risk for having a baby with a genetic disease, such as sickle-cell anemia, Tay-Sachs, or cystic fibrosis. When one studies genetic diseases, the most obvious forms of pathology, it is fairly easy to divide DNA sequences into two categories: a narrow category of functional or normal, and a broad category of dysfunctional or abnormal.

As we noted in Chapter 8, however, we know virtually nothing about the DNA sequences producing the range of normal genetic variation in our population: tall-medium-short stature; dark-medium-light complexion; coordination; spatial perception; limb proportions; body build; nose form; susceptibility to allergies; hairiness; or any other of the multitude of ordinary phenotypes we encounter every day. These are precisely what "sequencing the genome" would miss—the range of variation produced by the extensive genetic diversity in the gene pool.

By virtue of focusing strictly on pathologies, the Human Genome Project, as originally proposed, falls into precisely the same conceptual framework held by the eugenicists. This is the idea that there is a single normal state for a given phenotype, whose nature is self-evident, and against which any deviation must be judged. This is true only to the narrowest extent, in the study of medical pathology. Where the eugenicists fairly explicitly held all deviation (including social and moral) from a narrowly defined ideal to be genetically pathological, the assumptions of contemporary genome enthusiasts are instead more implicit in simply adopting the paradigm of medical pathology to molecular research.[18]

As its conceptual shortcomings became better-known, the original Human Genome Project mutated, and was reformulated with two modifications.[19] The first involves focusing on the construction of a "map" of the genome based on restriction sites and length polymorphisms. Again the primary goal is for genetic pathology: specific RFLPs that are consistently found associated with pathological syndromes will help to localize the site of the genetic defect responsible in the genome. Nevertheless, the groundwork may also be laid for actually coming to grips with the broad range of normal genetic diversity in the human gene pool.

The second augmentation to the Human Genome Project is the "Human Genome Diversity Project," in which an ancillary objective becomes the preservation of cells from diverse aboriginal populations of the world. The expressed goal is to ensure that future students of human genetics will be able to have access to exotic gene pools.[20] Here it is certainly admirable that the focus is specifically on variation in the human gene pool, though the focus is on differences among populations. The a priori knowledge that most human genetic variation is polymorphic, rather than polytypic, should make it more important to preserve many samples from relatively fewer groups, than to preserve few samples from many groups, if one wishes to study the general extent and nature of human genetic diversity.

The major goal in this effort, unfortunately, is thus also guided by an archaic idea: the establishment of the ultimate genetic phylogeny of human groups.[21] In pursuit of this objective, advocates are obliged to maintain that non-European human populations are generally "pure," and have been spared the vagaries of history, of contact, and of gene flow— assumptions that are certainly gratuitous.[22]

WHO IS RELATED TO WHOM?

The Human Genome Diversity Project has a fundamental rationale for focusing on polytypic exotic populations, rather than on polymorphic local ones. That goal is the reconstruction of biological history within the human species: to discern which human populations are most closely related to one another, and when they diverged from one another.

This is not a novel research program within biological anthropology. As we saw in Chapter 7, genetics has been a tool for reconstructing such relationships since the collection of the very first blood group data. And the biological history derived from genetic data has not proved to be perfectly reliable: in some cases genetic data have yielded absurd inferences, and in many others, such inferences are extensively contradictory. There are two main reasons for this.

First, it is not particularly clear just what genetic data are needed to reconstruct population history, or how to analyze them. In the case of the split of humans, chimpanzees, and gorillas from one another, the divergence of the three genera approximately 7 million years ago appears to have been so close in time that it is most reasonably rendered as a three-way split, or trichotomy. Different bits of genetic data link them pairwise in various combinations, and it appears as though the genetic data preserve a biological history sufficiently complex as to preclude the drawing of a neat, perfectly bifurcating, tree.[23] The same is true of branching within the human species.

Second, it is even more difficult to infer divergences *within* a species than among species. Reconstructing biological history above the species level subsumes the knowledge that species are reproductively isolated from one another, so that once a species splits into two, they can only diverge from that point on. But *below* the species level, when one reconstructs *population* histories, that assumption does not hold. Assuming that parallel evolution is fairly rare, species share derived characteristics for one reason: they inherited them from a recent common ancestor. Populations, on the other hand, may share derived characteristics for two reasons: a recent common ancestor, or recent genetic contact. After all, wherever human groups have met, what Cole Porter called "the urge to merge" has invariably expressed itself (Figure 9.8).[24]

Where the biological history of a population is known to have included significant amounts of gene flow, it can easily be documented: For example, one finds contemporary Mexicans to be broadly intermediate between Native Americans and Spaniards in many allele frequencies. Of course the degree of intermediacy is highly variable, but that knowledge is important for interpreting the genetic patterns encountered in that population. But what about populations whose histories are not so well known—such as those of most of the aboriginal inhabitants of the world?

Some generations ago, it was generally assumed that only Europeans were socially cosmopolitan, and other peoples were generally isolated and pristine—that Europeans had history, and others did not.[25] Anthropologists now appreciate the difficulty of that assumption: other cultures are not "frozen in time," and other peoples are not "completely isolated" from one another, except in very extreme cases. Some could have been considered isolated from Europe, but today virtually no populations are, either culturally or genetically.

Thus, the San peoples of South Africa, targeted at the top of the Human Genome Diversity Project's list of isolated and unmixed populations, are neither.[26] And once again, we find this to be an old error in genetics. Anthropologists pointed it out to geneticists studying the Navajo in 1950: in spite of extensive ethnohistorical documentation of intermarriage between the Ramah Navajo and the Walapai, Apache, Laguna, Yaqui, as well as Europeans, geneticists insisted on the "purity" of the group.[27]

> The happiest women, like the happiest nations, have no history.
>
> —George Eliot

There is a historical paradox in all of this. The geneticists desire to turn back the clock and reconstruct the relationships of the world's pop-

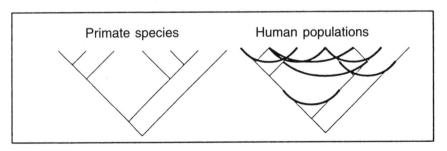

Figure 9.8.

ulutions as they imagine them to have been prior to Columbus. Under such a model, admixture is largely a recent nuisance. But of course, prior to Columbus there was Prince Henry the Navigator; prior to Prince Henry there was Marco Polo; prior to Marco Polo there were the Crusades; prior to the Crusades there were the Mongols and Huns; prior to the Mongols and Huns there were the Romans; prior to the Romans there was Alexander the Great. And those are just the European highlights.

Certainly there are inter-population differences to be analyzed and relationships to be reconstructed. But to approach these problems as if population contact suddenly began in 1492 or later, and to project a pseudo-history onto human population biology, are unlikely to be optimal intellectual strategies for studying the genetic variation in our species. We can, of course, genetically sample the human species of the 1990s, and attempt to study the composition of, and discern genetic similarities among, the populations that exist today.

The magnitude of difference between two gene pools can generally be attributed to four factors: (1) length of time of separation, (2) lack of subsequent genetic contact between them, (3) genetic response to local conditions or history, and (4) restricted size of the population. The first of these is the variable of interest. The second is the major confounding variable. The third involves the action of natural selection, and is difficult to assess other than anecdotally in most cases: the gene pools of Europe were almost certainly altered by the Black Death in the 14th century, and the gene pools of Native Americans by smallpox in the 16th century, though we don't really know how. The fourth is the action of genetic drift, under which the specific allele frequencies of small populations fluctuate and ultimately can reach zero. The South American populations that lack the Diego antigen, for example, can be presumed to have had it at one time, but to have lost it. It would seem more likely that such groups have been subject to the vagaries of genetic history— losing alleles from the gene pool due to population crashes or founder effects—than that they are not closely related to their neighbors.

To attribute all genetic patterns of similarity and difference in human populations to the first explanation, the time since divergence, is consequently highly unrealistic. Nevertheless it is not uncommonly encountered in the genetics literature—that a tree of genetic similarity of human populations represents the phylogenetic branching sequences of those populations.[28] The study of the genetic relationships of human populations, therefore, often makes some assumptions about the history of their gene pools that anthropologists regard as gratuitous.

This is why "phylogenetic" trees of human groups constructed on the basis of protein allele or DNA sequence similarity often vary extensively from one another. It is quite easy to extract the information that the Danish are more closely related to the Dutch than to the Iroquois, regardless of whether one's criteria are genetic or phenotypic. But for the details of biological history, it is unclear what kinds of genetic data are appropriate, what kinds of analyses are optimal, or whether "phylogeny" has a valid meaning in such a context. One can ask, after all, whether Cambodians are more closely related to Laotians or to Thais, but the forces that shaped the gene pool of Southeast Asia were operating long before the socio-political boundaries were erected, and independently of them. Consequently the three groups being compared are defined by highly arbitrary and non-biological criteria. To consider them as biological groups with a phylogeny to be discerned is to impose biological transcendence on historically ephemeral units. It is almost as misleading as asking whether lawyers are more closely related to architects or to accountants. One can always get genetic data and a tree from them, but the meaning of the tree may be elusive.

If we compare the trees generated by the two outstanding studies from the 1980s reconstructing human phylogeny from genetic data, those of Nei and Roychoudhury (1981) and of Cavalli-Sforza et al. (1988), we find them to be extraordinarily incompatible, in spite of the data and analytical methods being similar (Figure 9.9). Beginning with simply the details of the relationships among Asian and Oceanic peoples, the branching sequences of human populations the two studies yielded are quite different. And when we compare the relationships elucidated among the entire world's populations, we find the two genetic studies to be even more incompatible (Figure 9.10).[29] Indeed, about the only feature on which they agree is in having sub-Saharan African populations as the outgroup to the rest of the peoples of the world—though whether that feature would stand up with more complete sampling is certainly open to question.[30]

Genetics, thus, does not seem to resolve for us the nature of the large-scale relationships among populations. It can't, for that question is framed in an antiquated way, without

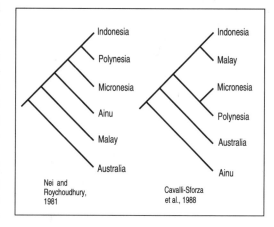

Figure 9.9. Relationships among populations of Asia and Oceania, according to two different genetic studies.

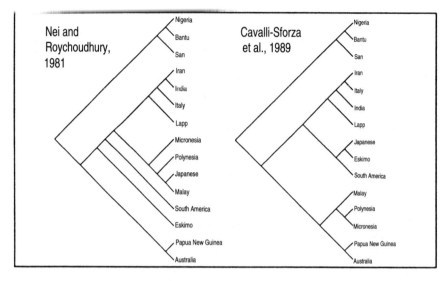

Figure 9.10. Relationships of world populations, according to two genetic analyses.

acknowledging the reticulated microevolutionary history of human populations. Like all genetic data, however, these studies have helped shed light on some specific questions. For example, it had been speculated on phenotypic grounds early in this century that the natives of Australia might be especially related to Europeans, as might the Ainu of Japan. But there seems to be no genetic support for those hypotheses; genetic data appear to falsify them.

So the intellectual payoff for decades of studying genetic differences among human populations has been real, but modest. The largest payoff has been, paradoxically, to come to grips with the limitations of such studies. First, genetic data do not seem to come up with that elusive single-digit number of basic human groups, any more so than phenotypic data do: apparently the human species simply does not come packaged that way. There is consequently no "correct" computer program or data set to yield a definitive answer—the question itself is "incorrect." Second, the great bulk of detectable genetic diversity in the human species is polymorphism; relatively little is polytypism. Third, differences in the extent of polymorphism across populations can be studied, but the microevolutionary processes at work seem to preclude them from revealing to us any but the most minor features of human biological history.

NOTES

1. An Affirmative Action compliance form from a major university paradoxically lists "Hispanic" among its choices under "race," then defines that category in terms of the countries of origin they considered to encompass it, "independent of race." On the 1990 U. S. Census form, "Is this person of Spanish/Hispanic origin" (Box 5) was a different question from "Race" (Box 4).

2. Gill (1986). See also Krogman (1962), Giles and Elliot (1962), Howells (1973), Rogers (1984), Shipman et al. (1985).

3. There are also senses in which Eastern European Jews, for example, are genetically similar to one another—for example in the elevated frequency of Tay-Sachs disease. Nevertheless, there is genetic evidence of extensive admixture. See Mourant et al. (1978), Livshits et al. (1991).

4. Kidd (1993).

5. Layrisse (1958), Layrisse and Wilbert (1961).

6. These are the three major alleles; a very rare one is also known. See Mourant et al. (1976).

7. Batzer et al. (1992), Perna et al. (1992), S. Tishkoff, personal communication.

8. The specific topology of the tree one gets from the restriction mapping data or the DNA sequence data has proven very contentious. In the present context, it is abundantly clear that nothing like "races" fall out of the tree. See Cann et al. (1987), Vigilant et al. (1991), Maddison (1991), Templeton (1992, 1993), Hedges et al. (1992), Maddison et al. (1992), Stoneking et al. (1992).

9. Horai and Hayasaka (1990), Merriwether et al. (1991).

10. Ferris et al. (1981). The same result appears to obtain for nuclear DNA as well, where it has been studied. See Ruano et al. (1992).

11. Wrichnik et al. (1987), Hertzberg et al. (1989), Schurr et al. (1990). The difference seems to be one copy or two of the 9 bp (base pair) sequence; apes have two, which indicates that the evolutionary event is a deletion. Presumably a duplication preceded it, in the more remote past.

12. Additionally, the "Asian" deletion allele has been found in some African populations (M. Stoneking, personal communication).

13. Scozzari et al. (1988), Schurr et al. (1990).

14. Wright (1931, 1932), Provine (1986), Crow (1990).

15. Dobzhansky (1959, 1963, 1970).

16. O'Brien et al. (1983, 1986). Of course, the physiological link between genetic uniformity and failure to breed well is missing. Caro and Laurenson (1994) argue that excessive homozygosity per se is not the major cause of the cheetah's problems, finding predation and other "environmental" causes to be of greater explanatory value.

17. Dulbecco (1986), Bodmer (1986), Watson (1990), Gilbert (1992).

18. Newmark (1986) attributes to Nobel laureate Walter Gilbert the thought "that the central purpose of a project designed to sequence the entire human genome is to provide a reference sequence against which variation can be measured by individual laboratories." For a discussion of ethical concerns about the

Human Genome Project, particularly in relation to the eugenics movement, see Resta (1992) and Garver and Garver (1994).

19. Walsh and Marks (1986), Lewin (1986), Davis et al. (1990), Lewontin (1992), Olson (1993), Hoffman (1994).

20. Weiss et al. (1993), Kidd et al. (1992).

21. Cavalli-Sforza et al. (1991), Cavalli-Sforza (1991), Bowcock et al. (1991), Roberts (1991a,b, 1992b). Kidd et al. (1993) and Weiss et al. (1992) emphasize other potential scientific benefits for this project.

22. Thus, Cavalli-Sforza et al. (1991:490) specify the need to study "isolated human populations," and specify a number of ethnic minorities. Roberts (1991:1614), in publicizing the issue, explains, "What each of these populations have in common is that each has been isolated and has only rarely—if ever—intermixed with its neighbors."

23. Marks (1992a), Rogers (1993).

24. See Roberts (1991a) for this thought, without attribution. The show was *Silk Stockings*, and the song was "Paris Loves Lovers," sung by Don Ameche on the stage, and Fred Astaire in the movie.

25. Wolf (1982).

26. Wilmsen (1989), Solway and Lee (1990). While these authors take different perspectives on the extent and nature of the San contact with other peoples, they are against the idea that the populations are pure and isolated.

27. Kluckhohn and Griffith (1950), p. 406.

28. See, for example, Cavalli-Sforza and Edwards (1964), Cavalli-Sforza et al. (1988), Nei and Roychoudhury (1981, 1993), all of whom take for granted the phylogenetic nature of the different trees they generate. Black (1991) discusses the problem of the incompatibility of genetic trees for South American Indian populations.

29. The apparent discrepancy in the relationships among Polynesians, Micronesians, and Malays in Nei and Roychoudhury (1981) reflects a difference between Figure 8 and Figure 10 in that paper. More recently, Nei and Roychoudhury (1993) contrasted the structure of trees of the human species constructed from the same genetic database with different computer programs. Their own (called neighbor-joining) reveals to them "five major groups" and "intermediate populations, ... apparently products of gene admixture of these major groups" (p. 937). This reflects archaic assumptions about human history and the structure of human variation. Sampling a few populations from five different parts of the world virtually guarantees identifying five rather discrete groups, as recognized by Bowcock et al. (1994).

30. There is some evidence, particularly from mitochondrial DNA, that sub-Saharan Africans are not so much the outgroup to other human populations, but *subsume* the diversity within all other populations. In other words, sub-Saharan Africans are paraphyletic. The studies cited in the text here are contrasting average allele frequencies for populations, and ignoring the diversity within each, though these data may well have phylogenetic implications.

10

The Adaptive Nature of Human Variation

Human groups often differ in adaptive ways, due to the action of natural selection. These adaptations include a range of responses to a broad suite of environmental challenges. Not all human differences are biologically adaptive, however. Human groups culturally define themselves in juxtaposition to other groups.

Polytypic variation in the human species is the variation we tend to focus on, in spite of the fact that it represents a fairly minor component of the biological diversity in our species. And the polytypic variation that does exist is structured not racially, but clinally. In other words, different populations do not seem to fall into a small number of large clumps, but seem to vary gradually over the map.

This pattern of variation is found regardless of whether one studies phenotypes in a low-tech manner, such as the proportion of people with light-colored eyes in Europe (Figure 10.1); or genotypes in a high-tech manner, such as the first principal component of a synthetic genetic map of Europe (Figure 10.2). In the former case we see a smooth North-South gradient, where the phenotype is concrete and the tabulation straightforward. In the latter case, where the scale is arbitrary and the analysis abstract, a similar pattern nevertheless emerges.

The appreciation of a clinal, rather than a racial, pattern of human variation is critical for reasons articulated in 1932 by Lancelot Hogben:

> Geneticists believe that anthropologists have decided what a race is. Ethnologists assume that their classifications embody principles which genetic science has proven to be correct. Politicians believe that their prejudices have the sanction of genetic laws and the findings of physical anthropology to sustain them. It is therefore of some importance to examine how far the concepts of race employed by the geneticist, the physical anthropologist, and the social philosopher correspond.[1]

Hogben went on to show that they do not correspond well at all. As we have seen, they showed little correspondence even among physical

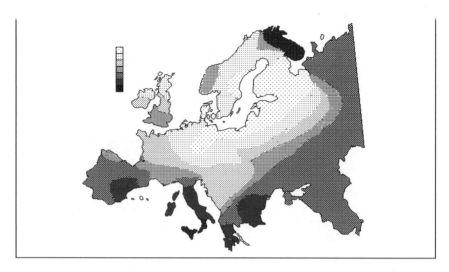

Figure 10.1.   Proportion of people with light-colored eyes. Top of the scale
    corresponds to less than 79%; 65–79%; 50–64%; 35–49%; 20–4%; 10–19%;
    1–9%. After Hulse (1963:328).

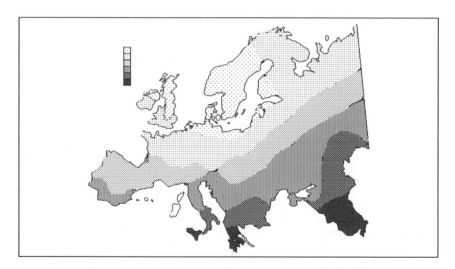

Figure 10.2.   First principal component of genetic variation in Europe,
    accounting for 28% of the total variation. After Cavalli-Sforza et al. (1993).

anthropologists alone. The biological pattern these scientists sought simply doesn't exist, as was expressed epigrammatically by Frank Livingstone: "There are no races, there are only clines."[2]

However, Livingstone overstated the case in a significant way. Races do exist, but they are social categories. The mistake is in relegating the social patterns to trivia—when they are of paramount importance in our daily lives. We sought a biological, scientific validation for the distinctions we made between "us" and "them." Those distinctions were, and are, very real and terribly important. If anything, it is the *biological* patterns that are trivial, at least to any ambitions we may harbor about carving up the human species into a small number of discrete and relatively homogeneous groups. If we study human biological variation, we are obliged to defer to its clinal nature.

The reason we find these patterns lies in the biological microevolutionary history of our species. Gradients are produced by the contact between populations that are slightly different genetically; and by the local adaptations of populations to the specific environmental conditions in which they exist. As environments change gradually over space, so too do the human adaptations to them.

The biological structure of the human species, then, is a dynamic structure reflecting both history and descent. We see the same kinds of patterns—gradations, rather than clusters and sharp discontinuities—over other regions of the world as well. We can trace a simple phenotype, variation in skin color, and a complex measure of genetic similarity and find the same kinds of patterns. Not the same pattern, obviously, but the same *kind* of pattern. Interpreting the distribution of these traits, then, becomes the focus of the study of human biological variation. This is a far more empirically based science than trying to discern what the basic races are. Even so, it is often vexing: the African cline in Figure 10.3 is thought to reflect primarily patterns of gene flow; and that in Figure 10.4 to reflect a basic adaptation to solar radiation. Whether this is indeed the case is more difficult to say.

PATTERNS OF GENE FLOW

The genetic differences that exist among populations have provided opportunities to study rates and proportions of interbreeding. Though human populations have probably always been in some degree of contact, history documents several extensive examples of biological contact between populations from widely different parts of the world.

As discussed in Chapter 6, social stratification often has an effect on gene flow: people of "mixed race" tend to be classified with the socially

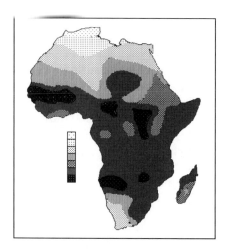
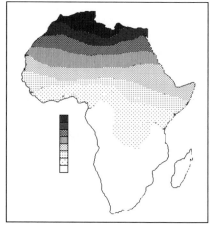

Figure 10.3.   First principal component, representing 35% of the
total variation, of a synthetic
genetic map of Africa. After
Cavalli-Sforza et al. (1993).

Figure 10.4.   Variation in skin color
across Africa, after Biasutti (1942).

lower group. Consequently, it is that group that typically expands to accommodate the gene flow. Where there are major differences in allele frequency between west Africans and western Europeans, it becomes possible to calculate the extent to which western European mixture has contributed to the gene pool of African Americans. It varies around the United States, but a general estimate seems to be about 22–25 percent of the African American gene pool deriving from western European admixture.[3]

Eugenicists, taking racial purity as a goal, were at great pains to demonstrate that race mixture was detrimental to the genetic endowment of the offspring. Race mixture was sufficiently threatening socially that its badness required a scientific, biological justification. And yet such evidence, when forthcoming, did not stand up well to scrutiny: it proved exceedingly difficult to show any negative biological effects of race mixture.[4] Not only that, but the "dysgenic" effects of interbreeding were difficult to justify theoretically; if anything, by analogy to domesticated species, the increase in heterozygosity in hybrid human offspring should make them fitter—whatever that might mean with respect to humans.

The commingling of human populations, however, is neither new nor rare. The pervasive presence of clines all over the world implies the contact of populations, and the biological history of Europe seems to be particularly well-explained by repeated and extensive waves of documented large-scale migration, in addition to the long-term gene flow occurring on a smaller scale.[5]

Given that one expects to find geographic clines as the result of both interbreeding and of local adaptation to gradually-varying environments, it becomes a problem to explain any particular cline. Implicit in an adequate explanation is a knowledge of both history and physiology: knowing something about the demographics of the population, and the function of the trait and its alternative forms. Specifically, the issue becomes: Is the trait adaptive, or is its geographical distribution merely the passive consequence of the movement of populations?

ADAPTATION

Adaptation is a particularly troublesome concept in evolution, because it (1) means two things, (2) arises in several ways, and (3) is not easy to identify when present. Nevertheless, many facets of human biological variation are plausibly regarded as having an adaptive nature, providing some benefits to the individuals possessing them, in specific environments.

The two things we mean by "adaptation" are, first, a feature providing a benefit over its alternatives to an individual in a particular environmental circumstance, and second, the process by which such a feature arises.

Adaptation—the process of adapting—itself subsumes four different processes. First, the "classical" definition: Adaptation occurs by natural selection, as a consistent bias in the survival and reproduction of individuals with particular genetic configurations. Here adaptation is a genetic process, and the identification of a feature as an adaptation implies that the differences between individuals with the feature and without the feature have different genetic bases. Second, the "facultative" definition: Adaptation occurs as the result of physiological or behavioral plasticity, via long- or short-term environmental stress on a relatively undifferentiated organism. Here, adaptation is a developmental process, the result of a physiological system with many possible endstates, and individuals with and without the feature may well be identical in their relevant genetic information. Third, adaptation occurs as the result of choices made by individuals in direct response to a particular situation. Here the adaptation is behavioral, but the actors' choice of behavior is one that directly enhances their welfare. And fourth, a particularly human mode of adaptation, in which a corporate decision is made that may sacrifice short-term benefits to individuals in the interest of long-term benefits to them as a collectivity. Though this is also a behavioral adaptation, it is one that is based on the unique abilities of humans to be foresighted, and on the cultural institutions that can compel individuals to take a short-term maladaptation for the long-term good of the group.[6]

In general, the first kind of adaptation produces morphological differences among groups of individuals; the second produces morphological differences between groups or among individuals within a group; the third produces behavioral variation within and between groups (and is the focus of much of contemporary human behavioral ecology); and the fourth produces behavioral differences among human groups.

Identifying particular features as adaptations—the other use of the term—implies a great deal of knowledge: the use of the feature, the origin of the feature, and the manner by which it arose. Often, however, this knowledge is incomplete, and one is obliged to be tentative about assuming that specific features are adaptations. The ambiguity stems from the fact that a particular biological structure has several uses, and it is consequently often unclear as to which use was "the" adaptive function for which the structure evolved.[7]

Take, for example, the human hand. It is capable of more fine-scale manipulation than an ape's hand (Figure 10.5). Humans use their hands to modify nature extensively, making tools. But did our hands evolve for that? Are our hands adaptations for tool-making? Humans also use their hands extensively during sexual activity, to stimulate their partners, unlike apes. How can we know whether the first or the second use of the human hand is the adaptive explanation? That would relegate one or the other of the uses to the status of "additional benefit" of the hand. In fact, how can we know that either is the explanation, and that the human hand didn't evolve for some other reason and that *both* are "additional benefits"?

Alternatively, the possibility must be entertained that a particular feature emerged for one reason and was elaborated for another. Contem-

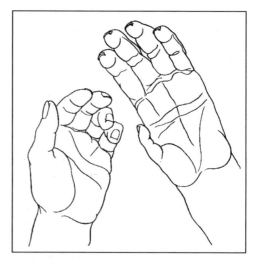

Figure 10.5.   Hand of a human and a chimpanzee, after Napier ([1980] 1993).

porary explanations for the evolution of feathers in birds, for example, center around their emergence for thermoregulation, and their elaboration for flight. Gould and Vrba[8] have sought to formalize these insecurities in the study of adaptation, by distinguishing features that evolved for a specific function (adaptations) from features that evolved for one thing, and were co-opted by another function, for which the feature turned out to be advantageous (exaptations).

Thus, although we acknowledge that features are adaptive, it is difficult to use that concept in a testable, "scientific" way. That something has adaptive value implies that it has a specific reason for existing, which is a comforting way of approaching the universe, but is not necessarily accurate. As mentioned in Chapter 2, the adaptive "story" is often constructed as quite literally that: a story, in a literary format, with plausibility as the main criterion for its validity.

There are several mechanisms by which a biological structure could come into existence *without* there being a particular reason for it. The first is simply the random process of genetic drift in evolution. While the high frequency of sickle-cell anemia within equatorial African populations is an adaptation to malaria (Figure 10.6), the high frequency of a disease called porphyria in white South Africans is a simple consequence of descent from one of the original white settlers in the early 18th century, Jacomijntje van Rooyen.[9] The distribution of sickle-globin is clinal and adaptive; that of porphyria is not.

Another non-adaptive explanation for biological structures was known to Darwin as the "correlation of parts." What, for example, is the cause of the form and structure of the human ring finger? Presumably, it is that whatever force operated to produce the form and structure of the other digits operated on that one as well. Likewise, the presence of nipples in men presumably has to do with the fact that they function in lactation in women, and men and women develop according to a common genetic program. In these cases, there may well be an adaptive significance for the feature in question, but it may not have involved selection specifically for that fea-

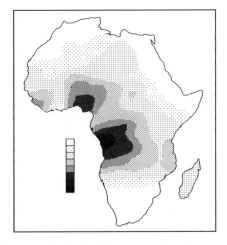

Figure 10.6. Frequency of the sickle-cell allele in Africa, ranging from zero to about 25%. The allele also is present in South Asia and around the Mediterranean.

ture. Rather, the feature must be understood with reference to its cognates or homologues.

One of the most common anthropological narratives concerns the origin of brow ridges, present in our ancestors, but not in our species. Why were they there? To protect bearers against blows on the noggin? To keep the sun out of their eyes? To absorb the chewing forces that arose from the large jaws? Or possibly for no reason at all—simply as a passive consequence of growing a fairly large face attached to a skull with a small frontal region of the brain? It is, quite simply, very difficult to tell.

Biology and anthropology have experienced several types of "mood swings" throughout the 20th century, on the issue of the explanatory power of adaptation. The scientific mood accompanying the post–World War II optimism produced a "hardening" of evolutionary theory, within which nearly all structures were studied by reference to their assumed adaptive value. An organism was considered something in which all was well, for it had been composed by natural selection to fit perfectly into its world.

> In the realm of Nature there is nothing purposeless, trivial, or unnecessary.
>
> —Maimonides

In human studies, this viewpoint was stridently challenged as early as 1963 by Sherwood Washburn, criticizing postulates of racial adaptations. Was the form of the nose adaptive? No, argued Washburn.[10] (Yes, had argued Carleton Coon.[11]) If one actually studies the form of the nose empirically, one finds rather a small relationship to geography or climate. One finds highly variable forms, and no evidence for the activity of natural selection taking a toll on offspring of individuals with wrongly shaped noses.

Consequently, the "adaptationism" of the 1950s and 1960s has given way to a modern eclecticism in explaining biological structures, more recently as a result of the criticisms of Lewontin and Gould in 1979. The modern view is formulated partly on the model developed by anthropologist Claude Lévi-Strauss to explain how myths develop in human societies. Myths, says Lévi-Strauss, do not come together in a single work, nor are they crafted precisely by refining every element. Rather, they are assembled from bits and pieces that seem to fit well together, and are jerry-rigged into the complete structure. In every generation certain elements are added, others tossed out, others reworked, by a mythmaker who functions as a *bricoleur*, or tinkerer.[12] This analogy was transferred from cultural to biological evolution in 1978 by the French molecular biologist François Jacob, whose widely-cited argument was that *biological* evolution is like tinkering, not like engineering.[13]

Acknowledging that organisms are not assembled by an all-wise creative hand of natural selection poses difficulties for the explanation of biological forms. We can no longer merely assume that features arose to fulfill the main function they currently have, nor that any of the functions they have is the main function, nor that they arose to fulfill specific functions in the first place. This is not to deny the role of natural selection, nor to deny the reality of adaptation, but to question the reasoning behind the assumption that anything is an adaptation without a specific test for it.

Ultimately, then, localized variations in human appearance need not be explicable as adaptation. We may be struck by the aboriginal Ainu men of Japan, who are far hairier of face and body than other Japanese; or by the Khoi (Hottentot) women of south Africa, with elongated labia and large fatty deposits on their buttocks, known as steatopygia. But whether these local differences are adaptive, or even functional, is difficult to discern; and they are generally rare. These local variations once fascinated anthropologists, when the field was centered on the rare and exotic.[14] But as it has generalized to studying the world at large and the world at present, anthropology has progressively come to see these features as largely trivial components of the overall biological picture of the human species.

GENETIC ADAPTATION

Genetic variation in the form of mutations is the basis of genetic adaptation. Mutations, however, are almost universally harmful to the organism in which they occur. The reason is simple: the genetic program has evolved over the eons, and random knocks to it are unlikely to improve it. Since most of the genome's DNA is not expressed phenotypically, most mutations will have no effect. Most mutations are therefore neutral mutations, but since they are unexpressed, they have little to do with adaptation.

Since humans live in structured populations, mutations with slightly deleterious effects can spread by the process of genetic drift, in spite of their effects. It is conceivable that in an appropriate genetic background, a mutation that is slightly harmful in one context may turn out to have beneficial effects in others, or in conjunction with other genes.

More important in a consideration of genetic adaptation, however, is our implicit view of normality. The classic view of mutations derives from the early Mendelians, and comes to us by virtue of its great successes in fruitfly genetics and medical genetics. Here, a pathological phenotype is isolated (say, "white eyes" or "cystic fibrosis"), and its genetic etiology is traced back to a particular mutation in a particular gene. We can call this the "pathology paradigm" because it focuses very specifically on pathological alterations to the phenotype, often leaving largely unanswered two central questions of students of evolution: the gene's *normal* function (to prevent cystic fibrosis?), and the genetic basis of the range of normal phenotypic variation in a population (such as height, body build, coordination, facial conformation, or various forms of intelligence).

Consequently, as noted in Chapter 8, the map of human genes translated into phenotypes is largely a map of genetic diseases. And for any particular gene, relatively little is known about what the gene actually *does*, aside from causing a disease when changed. Further, there is a pool of genetic variation "out there" in normal populations, whose phenotypic expression is obvious and variable, but whose genes are largely unexplored.

Mendelian genetics has had its most notable successes in the study of phenotypic pathology, but due to the complex physiology of gene action has contributed relatively little to the study of the normal range of human phenotypic variation. This was one of the major pitfalls of the ideology behind the eugenics movement: the narrow definition of normality, which implied that much of the deviation from it was pathological in nature.

To return to the genetic nature of adaptation, this kind of approach from Mendelian genetics led to a particular conception of the role of mutations in evolution. In evolutionary terms, each species was considered to have an optimal homozygous genotype for any specific genetic locus. A mutation would therefore be a rare pathological deviation from that optimal form. Theodosius Dobzhansky recognized the Platonism inherent in that conception of the nature of genetic variation: it assumed a single ideal best type of organism, against which all real organisms were degenerations. His proposal to counter this "classical" model is the "balance" model, in which the optimal genotype is often *heterozygous*.

Dobzhansky proposed that genetic variation is *necessary* for a population to thrive[15]. In other words, the population composed of the best-adapted individuals is one in which there is considerable genetic heterogeneity. Evolution, then, rather than being a simple transition from homozygote (A_1A_1) to homozygote (A_2A_2), is seen as a more complex transition within a heterogeneous gene pool. This provides a simple explanation of the well-known phenomena of "inbreeding depression" (inbreeding promotes homozygosity) and "hybrid vigor" (hybridization restores heterozygosity). Experiments on fruitflies in the 1950s and 1960s seem to bear this out well, as well as the discovery of extensive variation in enzyme forms in natural populations and more recent observations concerning the effect of the loss of genetic variability on the immune system.[16]

The major implication of Dobzhansky's "balance" model is the broadening of our conception of normality at the genetic level. Rather than there being a single optimal allele with a single optimal homozygous genotype, we are obliged to consider a variety of alleles at any particular locus, various combinations of which may afford optimal genotypes. Mutations, therefore, can (at least within limits) be constructive, as long as it is recalled that they function in a diploid organism, with complex physiological processes dictating the emergence of phenotypes from particular genotypic arrangements. This requires retreating from a narrow and restrictive conception of what is "normal" or "optimal," which is arguably one of the major conceptual revolutions of the latter half of this century, and a theme to which we will return.

HUMAN VARIATION AS PHENOTYPIC ADAPTATION

Differences among human populations in general appearance are due partly to adaptation to three basic environmental variables: heat, light, and oxygen. Body build tends to vary with heat, according to principles derived from geometry and physics. A sphere has the highest ratio of volume to surface area. The amount of exposed surface area is what determines the efficiency by which is dissipated (we huddle in the cold, and fan ourselves in the heat); and consequently a spherical object retains heat most efficiently and radiates it least efficiently, of any geometric solid. This seems to be the reason we find the world's stockiest people in the arctic, where thermal retention is at a premium, and the world's lankiest in east Africa, where thermal dissipation is at a premium (Figure 10.7).[17]

It can't be over-repeated that there is extensive variation within populations in body build, and that this is developmentally responsive to environmental variation to some extent, as migrant studies and changes in indigenous population through time have shown. Nevertheless, there are some generalizations that can be made on the basis of average differences among populations. Bergmann's rule, which is general among mammals and applies to humans, finds a mean annual temperature varying inversely with body mass—heavier people tending to be found in colder climates. Allen's rule likewise finds animals and people with longer limbs in warmer climates.

Populations living at extremely high altitudes have a number of physical specializations. Though studies of the Nepalese atop the Himalayas are slightly at variance with those of the Peruvians atop the Andes, it appears that the oxygen stress of these environments has caused the physical form of the residents to respond developmentally in adaptive ways.[18] These ways involve expansion of the lungs and broadening of the chest.

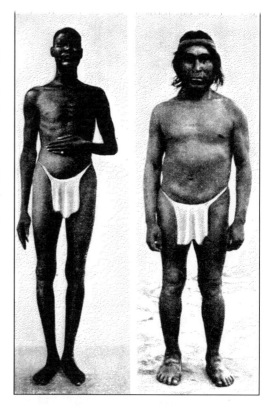

The human body responds facultatively as well to levels of ultraviolet light, by tanning. Populations also differ genetically in the number of melano-somes—pigment granules containing melanin—contained in skin cells called melanocytes. Because darkly-skinned people are widely distributed throughout the world, this is more likely the ancestral condition for the human species. Since sunlight is required to activate vitamin D in the human body, it is believed

Figure 10.7. People from different parts of the world often have different average body builds. After Biasutti (1958), retouched from the original (without loincloths) in Martin (1914).

that the migration of humans into higher latitudes led to stress from rickets, which is caused by a vitamin D deficiency. Depigmentation would then have been a genetic adaptation to lower levels of sunlight than ancestral humans would have been exposed to, shielding them from rickets in the new environment.[19]

The distribution of pigmentation in humans also follows a generalization known as Gloger's rule: that animals found in the wet tropics are darkest; in the desert are brown; and in the arctic are white. Though there is no satisfactory explanation for it, it seems to apply to the human species at the microevolutionary level.

NUTRITIONAL VARIATION

Human populations vary in their metabolic functions, and consequently in their reaction to certain foods. Though the biological details are not well understood, one of the most well-known modes of variation in our species lies in the ability of approximately 30 percent of the adults in the world to drink milk without ill effect. All human infants can digest milk, but in most members of most populations, the activity of an enzyme called lactase, which breaks down the sugar in milk (lactose), declines with age.

Populations are polymorphic for this (presumably genetic) trait, but the minority of people in whom lactase continues to be produced through adulthood are the descendants of dairying cultures. In most adults—Africans, Asians, Australians, and Native Americans—drinking milk produces bloating, flatulence, and diarrhea. In many Europeans, however, the child's capacity to digest milk remains throughout adulthood.

Failure to recognize this diversity led to embarrassing episodes in the 1950s and 1960s, as Eurocentric institutions sought to relieve hunger in developing countries by providing them with an excellent source of nutrition—milk. Alas, most of the people to whom it was provided were sickened by the milk, which adversely affected their attitudes toward the West.

Approximately 78 percent of "white" Americans can drink milk as adults, as compared with 45 percent of "black" Americans. Though the frequencies do differ among populations, it is polymorphic in all populations. If a nutritional advantage accrues as the result of having this allele, it is not clear why it persists as a polymorphism in Europeans, and has not been fixed by selection. Neighboring peoples who differ in their culture histories (dairying vs. non-dairying) differ in this trait, but it is not clear why the trait would be polymorphic even in people with no history of dairying.[20]

Another major adaptation to nutrition seems to be related to the ability to thrive on the high-carbohydrate diet characteristic of developed, industrial society. In many peoples, obesity and diabetes are far more common now than just a few generations ago, concordant with a major shift toward "Western" diets. This has been most noticeable in Native Americans of the southwest, and Polynesian and Micronesian groups.

One popular explanation for the rise of obesity and diabetes is known as the "thrifty genotype" hypothesis, which holds that some human populations have metabolic adaptations to environments in which food shortages are regular. These adaptations might enable them to extract calories from foods more efficiently; but when high-calorie foods are plentiful, the system "overloads," and the body continues to extract its nutrients efficiently, producing obesity and diabetes.[21]

UNIQUENESSES OF HUMAN ADAPTATION

The human species is noteworthy ethologically for a peculiar aspect of adaptive behavior. Other species behave in an adaptive manner—in other words, in ways that favor the transmission of their genes; and their behaviors can profitably be examined in such a light. In humans, however, cultural goals may arise that conflict with the biological imperative to reproduce. Thus, any specific human behavior may not be interpretable *a priori* as a strategy of reproduction.

Human societies, for example, have economic systems within which individuals strive consciously toward particular ends. That end may be the improvement and comfort of their own life, and having children may lead to a perceived conflict with that end. This phenomenon is demographically well documented in the affluent classes of industrialized nations, and is presumably the cause of the lower birth rate that caused such anguish to the eugenicists.[22] (When coupled with the assumption that the lower classes are constitutionally inferior, this indeed would be a reason for concern; on the assumption that the babies produced by the lower classes could actually be capable of running the country, there is nothing to worry about.)

In this sense, then, *not reproducing* is an adaptation to an aspect of the cultural system. It is one in which the perceived benefits to one's existence outweigh the liabilities of being a "genetic death."

Similarly, people often curb their reproduction for ideological reasons, rather than for economic reasons. Celibacy and abstinence have helped to provide the moral authority of asceticism in many societies over the span of human history. Thus, a different aspect of the cultural system can also produce a goal at seeming odds with the biological imperative. Again, *not reproducing* is an adaptation here to another aspect of the cultural system.

A third goal that conflicts in humans with the biological urge to reproduce is an outgrowth of the "new" function of sexual activity in the human species. While most primates are rarely sexually active, and only when the female gives a clear signal (behavioral, olfactory, or visual) that she is fertile, humans are somewhat different. By far, most human sexual activity occurs outside the context of reproductive effort.[23]

Obviously reproduction is impossible without sexual activity, though technical advancements in artificial insemination and *in vitro* fertilization vitiates even that strong a thought. However, unlike most other primates and especially our closest relatives, human evolution has produced a species that is sexually active far more frequently than is necessitated by reproduction. This being the case, it seems fruitful to seek another function that has emerged as primary in humans for sexuality. That function appears to be to permit an emotional intimacy or bonding between the two participants, and it appears to be as much a basic biological property in our human species as bipedalism.[24]

Consequently, it becomes less difficult to interpret the fact that in most human sexual activity, the participants are not trying to reproduce, are incapable of reproducing, or are actively hoping not to reproduce. It appears to be a fundamental aspect of human nature—the loosening of the bond between sex and reproduction, which are so much more intimately connected in other species. One interesting exception to this appears to occur in our close relatives the bonobos or pygmy chimpanzees, who are far more sexually active than common chimpanzees or other primates, and appear to "use" sex in socially creative, nonreproductive ways.[25]

It is nevertheless amusing occasionally to encounter "biological" explanations for human behavior that fail to take account of the biological idiosyncrasies of the human species. Two popular books by ornithologists have suggested that courtship rituals of male birds, displaying in order to have an opportunity to reproduce, are similar in nature to the "displays" of men in fast, cool cars.[26] Alas, a bit of field-work could easily disabuse any thoughtful biologist of this idea: most people using cars to attract sexual partners are hoping (usually very strongly) *not* to reproduce as a result of their display.

In a similar fashion, one can explain the prevalence of homosexuality as another consequence of the segregation of sexuality from reproduction in humans. It stands to reason that if one can enter with a member of the *opposite* sex into an intimate sexual relationship within which reproduction will not occur, one could alternatively enter into a similar relationship with a member of the *same* sex with identical results. Looked at in light of the emergence of specifically human behavioral properties in contrast to our primate background, homosexuality requires no more special explanation than typical heterosexuality.[27]

In the case of homosexuality, we see a third tradeoff against the procreative goal. In addition to culturally-based economic and ideological goals toward which humans can strive at the expense of reproducing, humans have evolved a goal of emotional fulfillment by personal intimacy through sexuality, that can also conflict with the reproductive goal.

These goals (reproductive, economic, ideological, and emotional) can often be concordant. What makes human behavior so interesting and so variable, however, is when the goals are *not* concordant, and humans are forced to make choices about their behaviors from the various goals they are given, and the opportunities they face. It is clearly impossible to regard any specific human behavior as oriented toward the goal of reproduction, in the absence of a great deal of other information. This is analogous to the reluctance on the part of biologists nowadays to assume that the first detectable function of a particular anatomical feature is "what it evolved for" or "what it is an adaptation to."

CULTURAL SELECTION

One of the other unique features of human evolution is the emergence of a coercive authority from culture. In other species, ethological theory dictates that the behavior of an individual animal must be somehow in its own best interests, or at least in the best interests of its genes. This is because there is no conceivable way for a behavior to emerge for the good of the group, if it is not in the best interests of the individual animal as well. In humans, however, the coercive nature of cultural institutions can make people take short-term losses in exchange for long-term gains; or take losses in exchange for gains to others in positions of power. (Biological evolution does not permit this, since the currency of biological evolution is offspring, and organisms that take short-term losses in offspring either don't make it to the long-term or are "swamped out" when they do.)

Another importance of culture in human biology and behavior, then, is that culture can itself be an adapting entity, against which the short-term best interests of individuals can be traded. Cultures thus can take on their own historical trajectories, external to the individuals who participate in them.[28] The individuals' general interests are usually concordant with those of the culture, but they need not be.

Cultural traits can spread as if they were favorable alleles, by an analog of natural selection that we can call cultural selection.[29] Cultural selection acts to produce cultural adaptations; but these may not be adaptations directly for the benefit of individuals. A culture trait spreads at the expense of alternatives because more actors adopt it. It can spread because it directly enhances the welfare of the actor (for example, widely used advances in health technology), but this welfare may not be reproductive, and may not even be real, only perceived (such as the use of horoscopes, or faith-healers).

The widespread use of VCRs for home entertainment provides a mundane example of some of these processes. Certainly they have trans-

formed our lives, but their effect on our reproductive capacities is dubious; rather their value seems to be (at best) to reduce stress by providing accessible relaxing entertainment. And early on, they came in two
forms: VHS and Beta. These were, for all intents and purposes, equiva-
lents, yet VHS has driven Beta to extinction. The spread of these cultural
elements was driven by cultural institutions (principally corporations),
which promoted a demand for them.

Thus, although it is a truism that humans use culture as the primary
means of adapting biologically—of extracting the necessities for survival
and reproduction—the complexities of human existence dictate a far
broader role for culture. Cultural evolution not only augments the survival and reproduction of human beings, but promotes the proliferation
of cultural forms themselves, and of the institutions creating them.

CULTURE AS A SOCIAL MARKER

Perhaps the most extraordinary thing about culture is the immense
diversity it has provided to human groups. Humans appear to be very
variable, but that appearance is deceptive, for our views of variation are
often impressionistic and unquantified. Where it has been undertaken,
the quantitative study of genetic diversity finds humans to be far *less*
variable than our closest relatives, the chimpanzees and gorillas.

We noted this finding in the context of mitochondrial DNA in Chapter 9. Complementing this finding, a short stretch of nuclear DNA from
chromosome 17 was isolated by Ruano et al. (1992). Finding it to be
absolutely invariant across a diverse sample of humans from all over the
world, they then studied 16 homologous chimpanzee sequences and
found *two* alleles; and twenty gorilla sequences, and found *four* alleles,
one of which was very divergent.

The implications are striking: in spite of the African apes being rele-
gated to small relict populations localized to the African tropics, and
humans having expanded all over the world, the former seem to be
more genetically diverse than the latter. It would seem as though the
apes are remnants of once-extensive gene pools that have been only
recently cut back; while humans have undergone a comprehensive
reduction in genetic diversity that has not been replenished by their
demographic expansion.

Indeed, the biological history of the human species seems to suggest
a continual pruning of the evolutionary tree. Paleoanthropological systematics, discussed in Chapter 2, reveals that the human lineage 2 million years ago consisted of three genera: *Homo, Paranthropus,* and *Aus-*

tralopithecus. One million years ago, there was but a single genus, *Homo*; by 200,000 years ago, a single species, *Homo sapiens*; and now a single subspecies, *Homo sapiens sapiens*. The reason for this loss of taxonomic diversity is unclear, but it is certainly correlated with the elaboration of the technological aspect of culture as a means of adapting.

Paralleling this reduction in biological diversity, however, is a concomitant elaboration of a new dimension of diversity, a uniquely human dimension, a cultural di-

> No distinction shall be made between Trojan and Tyrean.
>
> —Vergil, The Aeneid

mension. Human groups, that is, culturally define themselves in juxtaposition to other groups. That self-definition is in the manner of speech, dress, customs, and appearance, and serves to compensate in some measure for the biological homogeneity in our species.

The predominant manner in which human groups vary from one another, indeed, is cultural. The great cultural variation within our species augments the biological differences between populations; indeed, it *swamps* the biological differences among populations. Like the recognition signals that identify members of particular species to one another, humans (lacking the biological differentiation of species, or even of subspecies) identify themselves as group members culturally. We identify ourselves as members of a particular culture in the way we dress, the way we decorate ourselves, the values we hold, and of course in the language we speak. Mating patterns in humans are very strongly constrained by attributes we perceive as similar in ourselves and our mates, and most of those criteria are cultural in their basis.

Group identification is certainly very fundamental to the human psyche. Our general inability to distinguish ourselves from neighboring populations biologically seems to have posed a problem that, like other solutions to human problems, was ultimately answered culturally. In this fundamental sense, then, culture has largely replaced biology as a manner of signaling membership in particular populations. Biological differences among populations exist in the human species, but these are generally subtle and continuous in nature; however, the cultural differences between adjacent populations are often very discrete.

NOTES

1. Hogben (1932:122).
2. Livingstone (1962:279).
3. Reed (1969), Lewontin (1991), Chakraborty et al. (1992).

4. Castle (1926), Herskovits (1927), Pearson (1930), Shapiro (1961), Provine (1973).

5. Sokal (1991a,b).

6. See Harrison (1993), Dunbar (1993), Morphy (1993). Paradoxically, in their zeal to study humans "just like other animals," one radical school of sociobiology in the 1980s tried to ignore the unique aspects of the fourth kind of adaptation. This would require seeing all aspects of human behavior as specifically directed to maximize reproductive fitness at the individual level (Betzig et al. 1988; Ruse 1988). It is, however, somewhat unbiological to try and explain human behavior without recourse to the most significant autapomorphy in the human genetic program: culture. Mayr (1988:79) notes as well that such "group selection" indeed operates uniquely on humans.

7. Williams (1966), Lewontin (1978), Krimbas (1984).

8. Gould and Vrba (1982).

9. Dean (1971).

10. Washburn (1963).

11. Coon (1962).

12. Lévi-Strauss (1962 [1966]).

13. Jacob (1978).

14. In the classic literature of the exotic, male genitalia attracted far less attention than female (though see Coon 1965:112–13, 153). On the Hottentot *tablier*, see Schiebinger (1993). Steatopygia is also known in other populations, such as the Andaman Islands (Coon 1965). This interest in buttocks was partly fueled by the apparent similarity of steatopygic women to the figurines found in Upper Pleistocene sites in Europe—the "Venus" figurines—although steatopygia is unknown in Europeans. See Bahn and Vertut (1988), p. 138.

15. Dobzhansky (1955), Beatty (1987).

16. Wallace (1958), Lewontin and Hubby (1966), Harris (1966), Ayala (1969), O'Brien et al. (1989), Black (1992).

17. Newman (1953), Roberts (1953), Bogin (1988).

18. Frisancho (1975), Baker and Little (1976).

19. Loomis (1967), Quevedo et al. (1985), Wood and Bladon (1985).

20. McCracken (1971), Kretchmer (1972), Harrison (1975), Flatz (1987), Saavedra and Perman (1991).

21. Neel (1962), Knowler et al. (1983), Weiss et al. (1984), Stinson (1992).

22. Coale (1983), Westoff (1986), Coleman (1990).

23. Nonreproductive sexuality is, of course, present among the primates to various extents—the human difference is quantitative, not qualitative. See Hrdy and Whitten (1987).

24. Related to this evolutionary novelty is eroticism. Our close relatives spend far less time engaged in any given copulatory bout, and far less time involved in tactile exploration of their partners' bodies. It is not clear that "erogenous zones" exist in our ape relatives, and the evolution of erogeny in humans can be conceived as both reinforcing the pair-bond, and antagonistic to it, depending upon the context.

25. De Waal (1989).

26. Barash (1979:78) invokes a Ferrari, while Diamond (1992, p. 175) invokes a Porsche.

27. Those who attempt to "explain" homosexuality are often at pains to define it adequately. Clearly human sociosexual behavior defies simple dichotomization. See Chapter 13.

28. Kroeber (1917), White (1949).

29. Cavalli-Sforza and Feldman (1981), Boyd and Richerson (1985), Rindos (1985), O'Brien and Holland (1990), Durham (1991).

11

Health and Human Populations

Disease has been a major stressor in human populations, and accounts for major aspects of global demography. Culture strongly influences patterns of disease in human groups. Paradoxically, we find culture affecting biological variables more than vice-versa.

One of the traditional assumptions in the discourse of the relation between culture and human biology is that somehow biology is an independent variable, affecting cultural patterns without being in turn affected by them. This consequently often boils down to assertions of an absurd nature: for example, that the monogamous social relations of one culture might be the result of ingrained biological propensities distinct from those of polygamous societies. Several generations of studies of immigrants and their subsequent acculturation have shown that there is nothing demonstrably ingrained in the biology of one group of people that suits them to select one particular cultural form over another.[1] Eastern European Jews over the last millennium, for example, have shown an equal facility for conversing in Yiddish, Hebrew, or English, depending upon where and when they have been raised. Likewise, transitions between systems of marriage seem to occur with comparable facility, but without concomitant alterations in the gene pool.

There is indeed a strong relationship between biology and culture, but the causal arrow usually points in the other direction. Cultural forms strongly affect the general biology of populations. Sometimes the biological parameter affected can be the gene pool directly, but often the biology is more broadly the biology of health and demography, with more somewhat indirect consequences for the composition of the gene pool. In any event, the effects to the gene pool lie not in any "genes for behaviors," but rather in the very gross structure of the gene pool.

DEMOGRAPHIC TRANSITIONS

Eugenics began with a very simple observational fact: the lower classes were out-reproducing the upper classes. In coarse Darwinian terms, this implied (by any standard of measurement) a higher biological fitness on the part of the lower classes. That is, after all, what evolutionary biologists mean by "fitness": the ability to pass genetic information into subsequent generations. And the lower classes seemed to be doing it better than the upper classes.

This conflicted with a very simple observational factoid: that the lower classes were constitutionally inferior to the upper classes. Inferiority naturally implied lower fitness in the technical deterministic Darwinian sense, so the fact that they were proliferating implied a subversion of natural law. The desire to curb their fertility can therefore be seen as an attempt to bring demographic realities into line with biological prejudices.

But what is really so bad for the country, or for civilization, about poor people having more babies?[2] If the babies are destined constitutionally to be impoverished or criminals, clearly that will place more paupers and criminals on the street—thus, the idealistic goals of eugenics and the ideology of hereditarianism came to reinforce each other. If that destiny can be altered, however, then the problem can be localized to the personal and familial spheres, rather than the social or national. The problem then would become, How do we get people to decide to reproduce less, and how do we recognize and develop the talents of whatever children are born?

Demography, the study of the history of human populations, reveals a number of regularities in the relationship between economics and reproduction. It is a truism that in general people behave in ways that accord with their best interests (or at least with their perceptions of those best interests). Reproductive behaviors are also governed to a large extent by economic considerations—or more generally by cultural concerns. In other words, people make reproductive decisions—often not consciously, but in accordance with the norms, or what they are "supposed" to do—that are economically beneficial.

In the history of our culture, and in other cultures throughout the world, we encounter three broad-based demographic regimes. Hunter-gatherers, who live by foraging, in largely egalitarian societies with little in the way of technology or private property, tend to live at low population densities. This is as true of contemporary foragers as of prehistoric foragers. The underlying rationale is simple: hunting and gathering only works efficiently if there aren't too many other people hunting and gathering around you.[3]

Somewhat stereotypically, as there is always variation in the actual data, the forager lifestyle involves a balance between fertility and mortality rates. One does not encounter rapidly expanding foraging populations, either now or in the archaeological record. The reason seems to be that children are a burden on women who are actively engaged in productive subsistence activities, gathering food. Thus the best known of these groups, the !Kung San of the Kalahari desert, tend to have about four to five children per lifetime, spaced about four years apart, and aided by the contraceptive value of on-demand breast-feeding.[4]

People who grow their own food, however, have very different demographic requirements. Again, there is a wide range of variation, especially in the kinds of agriculture-based societies encountered ethnographically, but some generalizations can be made. Although foragers are not tied to a specific tract of land, agriculturists are. They are obliged to settle there, to plant and harvest at the appropriate times, and to defend that land.

For this mode of subsistence, one does not need to locate the appropriate foods, as a forager must; rather, one must gain access to good land, protect it against usurpers, and mobilize a large amount of labor at the appropriate time of year. This is accomplished in two ways: by the development of extensive networks of kin ties, and by having more babies.[5]

A shift from a foraging subsistence to food production is an economic revolution of the highest magnitude, and it occurred independently in populations on several different continents. Though there are local variations, there is a demographic uniformity: settled, agrarian societies are invariably accompanied by an explosive rise in population.[6]

Again there is a range of variation, but societies based on intensive agriculture tend to have women in roles relating more to reproduction than to production: from resource acquisition to preparation; and from economic to domestic roles.

> This has always been a man's world, and none of the reasons hitherto brought forward in explanation of this fact has seemed adequate.
>
> —Simone de Beauvoir

(This appears to have a major effect on the status of women relative to men in agrarian societies, but for present purposes we will focus only on the demographic considerations.) The shift from a stable foraging population to a rapidly expanding agrarian population is a demographic transition, and in terms of the history of our own society, it was the *first* demographic transition.[7]

The demographics of an agrarian society are those which color our perceptions of our own history, for they reflect the values of our own society as they existed until the late 19th to early 20th century. That was the time when the economic trend that began with industrialism and culminated with modernization began to take hold, and to transform once again the structure of the population.

As women enter the work force in larger numbers, they make decisions to curb their reproductive output—to devote more effort to production than to reproduction—and technology affords them an ever-more-efficient means of doing so. This general decline in fertility is matched by a decrease in mortality, such that the fewer children who are born all stand a good chance of surviving to reproductive age themselves. This kind of population is characteristic of contemporary "first-world" countries, and is the result of a *second* demographic transition.[8]

It is not equivalent across all social classes, however. The less-educated tend to retain older value sets, and often lack the opportunities for women to enter the work force in other than low-paying, low-status jobs. The result is a lag between the upper and lower classes in the completion of this demographic transition.

In England, where class and "race" are far less intertwined than in America, demographic studies have shown this trend through time and across economic strata quite clearly. Women in the lower economic strata tend to get married at a younger age, tend to have their first child at a younger age, tend to have more children, and to have them earlier in life, than do their counterparts in higher economic strata. Over the last two decades, the transition has continued, affecting all social classes: lower classes are still reproducing more than higher classes, but less than lower classes did three decades ago.

Demographers find that the most predictive variable in these data is education. Educated women (who tend to be preferentially located in the upper classes) marry later and have fewer children than uneducated women. The reason seems to be that in general children are perceived as a disadvantage to a career. By contrast, women in the less educated classes tend to come from socially conservative homes, and to be less likely to talk candidly about sexuality or to plan their family. This also correlates with a

> He that hath wife and children hath given hostage to fortune; for they are impediments to great enterprises, either of virtue or mischief.
>
> —Francis Bacon

general ignorance or disapproval of contraceptive options. More importantly, however, people of the lower economic classes tend to have less faith in their ability to control their own lives, less concern for the future, and little personal ambition. Better-educated women tend to plan their lives more carefully, use more effective means of birth control, and have fewer unwanted or unplanned babies.[9]

DEMOGRAPHY VERSUS EUGENICS

These demographic trends provide a different, and more powerful, solution to the problem perceived by the eugenicists. Given that there is a large difference in fertility in America across ethnic groups, what can be done about it?

First, we can recognize that the fertility difference is not "racial" in nature, but economic; it is "racial" only insofar as "race" correlates with social class in America.

Second, since the cause is economic in nature, the solution must be economic too. If the problem is not in any real sense a biological problem, seeking a biological solution makes little sense. The obvious solution to the problem of differential fertility across classes is to *lower or remove barriers to upward mobility*. As people become upwardly mobile, their fertility declines—not by magic, but by the normal propensity of groups of people to try and make their lives easier.

Third, this is predicated on the availability of education, as a means of access into the middle class, and contraceptive options to implement the choices people make.

Fourth, and most explicitly in contrast to the assumptions of eugenics, is the issue of the "decline of civilization" as a result of the breeding of the lower classes. If the upper classes reproduce less, where will the gifted people to run the country come from? Obviously, from the upwardly mobile and prolific lower classes. Talented people are always appearing in all populations; the key is to *identify and cultivate those talented individuals regardless of the population from which they derive*. The supply of human resources—of people with the potential for excellence—is for all intents and purposes limitless; the failure of the 20th century has been an inability to tap into it, and (as in the case of the eugenicists) even to deny it exists.

ECONOMICS AND BIOLOGY

The most fundamental lesson to be learned from the study of demography is that where culture and biology meet, it is cultural forms that affect biological variables far more than vice-versa. Foraging, agrarian, and industrialized societies, separated by two demographic transitions, are accompanied by more biological differences than simply birth rates.

The most obvious biological consequence of subsistence economy is the pattern of health and disease encountered across various societies. Foragers live at low population densities with simple technologies, and consequently are generally beset by endemic disease—those which are always present in the environment. By contrast, epidemic diseases, which are transmitted from host to host, generally require a high population density to pose a major threat to the group. Without a high population density to sustain it, an epidemic can run its course in a short period of time, without threatening much of the population.

After a demographic transition resulting in large urban populations subsisting on an agricultural base, epidemic diseases pose a far greater threat to health. Leprosy, smallpox, syphilis, bubonic plague, tuberculosis, and typhus were all well-known major scourges in the pre-modern world. Several of these entered the urban centers of Mesoamerica along with the Spanish conquistadors, and rapidly decimated the populations.[10] Recently it has been suggested that indigenous populations may have been particularly sensitive to infections, due to a relative lack of genetic variation for antibody genes.[11] Whether or not this turns out to be true, the epidemics were able to spread because of the population densities associated with agriculturally-based societies.

The economic transition to agriculture takes another toll on human health. Though contemporary foragers have been pushed to the most marginal habitats on earth, those living several millennia ago were able to exploit richer, more temperate environments. And though we somewhat ethnocentrically used to model the adoption of agriculture as a move toward obvious improvement in the lives of foragers, current views on the problem are quite different. In both the Old World and the New World, the earliest adoption of agriculture is generally accompanied by what can only be interpreted as a *decline* in the quality of life. This decline is detectable by comparing the osteological remains of societies immediately pre- and post-agriculture, and finding smaller stature, greater indicators of nutritional stress, less resistance to disease as a result of nutritional stress, and generally shorter lives.

The earliest agriculturists, far from having made a brilliant discovery about how to grow their own food, appear to have been forced to adopt this subsistence strategy in the face of prospects worse than those which they were obliged to endure in adopting it. What were they enduring under agriculture? In the first place, a narrower range of foods. While hunter-gatherers subsist on a wide range of plants and animals, agriculturists rely on large quantities of less diverse foods. At the beginning of agriculture, this seems to have resulted in nutritional imbalances; for example, people relying heavily on corn seem to have had deficiencies of the amino acids lysine and tryptophan and the minerals zinc and iron. Second, a drought or blight may cause a foraging group to move on to another area and other foods, while an agricultural group, tied to its land and narrower diet, would suffer catastrophically. And third, agriculture seems to have brought harder labor to populations in the processing and preparation of foods.[12]

Among contemporary industrialized or modernized societies, advances in both sanitation and medical technology have diminished the toll taken by infectious disease. New diseases such as AIDS still pose general threats, but the overall effect on public health from infectious diseases in modernized societies is far less than in agrarian societies. Our high-carbohydrate, low-fiber diets, however, combined with lowered amounts of exercise and higher stress levels, make us more susceptible to heart problems, digestive disorders, and obesity.[13]

THE CULTURAL NATURE OF DISEASE

Not only are the kinds of diseases that affect human societies affected by culture, but the transmission of diseases is also strongly affected by cultural factors.[14] It was unclear whether kuru, a fatal degenerative disease of the nervous system affecting only the inhabitants of part of New Guinea, was hereditary or infectious. In the late 1950s and early 1960s, D. Carleton Gadjusek was able to establish that the disease was caused by a hitherto unknown kind of infectious agent, a slow-acting virus (now called a *prion*). It was acquired during mortuary practices, in which relatives of the deceased handled (and possibly consumed) the brain of the victim. In so doing, they ingested the kuru virus, which lay dormant for years, possibly decades, before producing the tremors, twitching, and dementia ultimately associated with it. Abolition of the funeral rites led to eradication of the disease.[15]

As kuru can be considered a disease whose spread was caused by the rituals of these New Guinea peoples, other diseases have associations with more familiar cultural forms. Polio, for example, does not seem to have been a paralytic scourge until the introduction of modern sanitary standards. When children played in what would now be considered unacceptably unsanitary surroundings, they often encountered disease-infested human excrement, exposure to which provided a "natural" form of immunization to children. Polio seems to have become a menace only after higher standards removed excrement from the children's environment, thus shielding them from immunization, and increasing the risk of contracting the paralytic disease at a later age.[16]

Likewise, tuberculosis, known from Neolithic Denmark and pre-Columbian America, is always associated with a high population density. Though the discovery in 1882 of the bacillus causing tuberculosis was a medical breakthrough, it had little effect on the death rate from the disease. The disease had been declining for a long time, because of other general health care measures.[17]

A classic case of the influence of culture on biology is that of the bubonic plague, which ravaged Europe in the 14th century, with outbreaks through the 17th century. Like kuru, it could not be cured (until the discovery of antibiotics), but ultimately only controlled. It is estimated that between 1349 and 1352, as the disease swept from Asia across Europe, some 20 million people succumbed to it, possibly onefourth of the population of Europe. Bubonic plague has been implicated in several turning points of European history: for example, in undermining serfdom, by making labor scarce and giving the impetus to wage labor; and in undermining the authority of the Church, which was powerless to deal with it.

The disease is caused ultimately by a bacterium (*Yersinia pestis*) whose primary host is a flea, whose own host is a rat. When obliged or inclined, the flea will bite a human rather than a rat, and transmit the disease to the human. After a two-week incubation, the victim develops large painful swellings known as buboes in the lymph nodes of the armpits, neck, and groin. Three days later, accompanied by extremely high fever, these burst under the skin, creating large black splotches. The cycle can then repeat itself, if the victim lives that long.

The spread of this disease, significantly, is predicated on the casual coexistence of fleas, rats, and humans. In other words, standards of public health and general sanitation were a prime requisite for the Black Death's European pandemic. Further, it was strongly tied to the development of the shipping industry: ships were particularly rat-infested, and promoted the spread of the disease from the port of Constantinople to the trading centers of Italy, and then outward through Europe. The presence of urban centers of high population density obviously promoted the transmission of the disease once it arrived, and the absence of a germ theory of disease at the time made it impossible to conceive of an effective treatment.

Ultimately, the ways in which European societies dealt with bubonic plague was over many decades to shift the emphasis to more effective prevention from ineffective treatment (usually ostracism, quarantine, and flight: ostracism forced victims to hide their condition as long as possible; quarantine condemned entire families and villages; and flight generally helped bring the rats and fleas to new populations). Higher standards of cleanliness, public sanitation, and the discouraging of circumstances promoting primary infections all worked over the ensuing centuries to prevent breakouts in England after the 1660s. At Marseilles, an outbreak in 1720 failed to spread beyond southern France, the last major outbreak of the Plague in Europe.[18]

In sum, the profound difference in the kinds of diseases that threaten the health of foraging and urban peoples makes it clear that cultural forms have had a major impact upon biological parameters. The diseases themselves have secondarily acted as selective factors on the gene pool.

ETHNIC DISEASES

Diseases whose spread is augmented by cultural factors can elicit a microevolutionary response from a population. In irrigated (and consequently mosquito-infested) malarial areas of Africa and to a lesser extent EurAsia, the sickle-cell allele is relatively common because of the defense it affords heterozygotes; likewise with beta-thalassemia and G6PD (glucose-6-phosphate dehydrogenase) deficiency in Africa, the Mediterranean, and the Near East, and with alpha-thalassemia in southeast Asia.[19]

We know little of the physiology of disease resistance, or of which alleles of what genes promote it. Beyond the case of malaria, indeed, most connections between genes and contagious diseases are largely conjectural.[20] Nevertheless, different human groups have been subject to different kinds of stresses from diseases, and also have unique allele frequencies, often of alleles that are rare elsewhere. As we have already noted, cases such as porphyria in white South Africans and Ellis–van Creveld Syndrome in the Pennsylvania Amish are readily explained as non-adaptive consequences of founder effects—all the copies of the allele presently in those populations are descendants of a single DNA sequence in a common ancestor several generations back.[21]

On the other hand, the allele for Tay-Sachs disease, a lethal defect of brain lipid metabolism in young children, is about 10 times more common in Jews of eastern European (Ashkenazi) ancestry than in other populations. Two lines of evidence suggest that this is *not* a result of founder effect. First, other genetic defects of brain lipid metabolism, notably Gaucher's disease, are also more common among Ashkenazi Jews than among other populations. While it is not unreasonable to expect a disease allele to have an elevated frequency as a result of random processes, it *is* unreasonable to expect several rare alleles involving the same metabolic system to be elevated by random processes. Second, the Tay-Sachs allele is not a single allele: the disease is actually genetically heterogeneous, for there are several different mutations to Tay-Sachs' disease, which are all present in Ashkenazi Jews. That in turn suggests that it is not caused by a single DNA stretch that has been passed on to a large number of descendants, but rather by several different DNA sequences having similar effects, which have all been independently elevated in the target population because of their effects—presumably favorable effects in heterozygous form.[22]

What factors might have promoted the fitness of the Tay-Sachs' heterozygote? The evidence is slim, but tuberculosis, which has traditionally been a scourge in overcrowded urban ghettos, has been suggested as a selective force, like malaria in west Africa. If true, then the socio-cultural factors driving the spread of tuberculosis (for the emergence of overcrowded ghettos is the result of social and cultural forces) may have had a major effect upon the gene pool of the Ashkenazi Jews.[23]

Even more difficult to explain is the prevalence of cystic fibrosis, for which about 1 in 25 people of northern European ancestry is a carrier, about five to ten times as large a proportion as found among Asians and Africans. Over 20 different fairly common alleles of the disease are known in Europeans. One of those mutations, known as ΔF508 (a deletion of phenylalanine, the 508th amino acid in the protein chain), accounts for about 2/3 of the cystic fibrosis alleles found. Thus, fully 1/3 are different alleles that have independently attained elevated frequencies.

Does cystic fibrosis confer resistance for some infectious disease upon heterozygotes? All the "biggies" have been proposed as the selective factor—tuberculosis, cholera, typhus, bubonic plague, influenza, malaria, and syphilis—but no convincing evidence has yet been adduced in support of any of them.[24] Nevertheless, the frequency and distribution of the disease suggest an explanation in the deterministic processes of heterozygote advantage, rather than in the stochastic processes of founder effect.

To the extent, then, that the gene pools of ethnic groups often differ by virtue of the diseases they harbor, it is likely that much of this genetic variation is due to historical-cultural factors. Of course, infectious disease is no longer one of the leading causes of death in modern society, as it was in earlier generations. But it need hardly be pointed out that the leading causes of death in industrialized societies—heart disease and cancer—are themselves mediated by cultural factors: diet, smoking, exercise, pollution, and stress.[25] Indeed, the distribution of high blood pressure and hypertension has striking socio-cultural associations. It is more prevalent in urban industrial than in traditional societies (and highest in societies undergoing economic "modernization"); and more prevalent in African-Americans than in whites. Between-group variation is largely accounted by four variables—dietary salt intake, body fat, physical activity level, and stress—while within-group variation appears to have a significant genetic component.[26]

The fact that different diseases are more strongly or weakly associated with different populations is knowledge of obvious value in making a medical diagnosis. Diagnoses are ultimately individual, but knowing the quantitative differences among populations can increase the chances of getting it right. This is true whether or not the risk of disease is known to be genetic: knowing what you are at risk for by virtue of circumstances or style of life can assist not only in diagnosis, but in prevention.[27] Some conditions with hereditary bases can be treated by altering the diet (for example, diabetes, phenylketonuria, or lactose intolerance). Avoiding other threats to health may involve more substantial behavioral modifications, by knowing what kinds of behaviors are situationally risky: for example, avoiding fatty foods if you are at risk for high blood pressure; avoiding alcohol if you are unable to use it in moderation; or simply taking appropriate precautions in the course of your occupation—electrician, coal miner, athlete, or biochemist.

CULTURE AND BIOLOGY: AIDS

Like bubonic plague in the 14th century, AIDS in the 20th century has been incurable, contagious, and augmented in its spread by a convergence of cultural factors. In this case, promiscuity and intravenous drug abuse have been two major factors involved in the spread of the disease. Though the specific course of the disease is unclear, AIDS is characterized by a deterioration of the immune system apparently triggered over the course of several years by a virus known as HIV. There appear to be three ways of contracting the disease: by being born to a mother with the disease, by sexual contact, and by the transfer of blood. Though other ways of contracting it are conceivable, they are about as unlikely as being struck by lightning.

Since the earliest cases in America were promiscuous gay men and intravenous drug users, the disease has occurred disproportionately in men and minorities. In Africa, however, where the disease has been spread by heterosexual prostitutes, men and women are stricken in roughly equal proportions. Its spread has been facilitated by the cosmopolitanism of modern life and by the ease of commercial travel.[28]

Like the bubonic plague, AIDS seems far easier to prevent than to cure. The key is again cultural: understanding the behaviors that play a large part in promoting the spread of the disease, and modifying them. There is another analogy to be drawn with a relatively recent epidemic in our cultural history, which sheds more light on the appropriateness of our responses. That epidemic was syphilis, and its transmission was venereal, like AIDS.

Being venereal, syphilis was actually protected by the sexual taboos in our society. Since neither the disease nor its manner of transmission could be discussed in polite company, misinformation about it spread far more extensively than information. For example, a book called *Racial Hygiene*, by a bacteriologist promoting the scientific betterment of society, explained to readers in 1929 that syphilis was a racial poison:

> The germ can be transmitted by sexual contact but also by kissing, by drinking from a common cup, by use of a common towel or any toilet article which remains moist between contacts. . . . We are in very little danger from the disease provided we are sexually moral, avoid close personal contact with others, and avoid contact with such things as may have been in recent intimate contact with others and have remained moist since the contact.[29]

In some circles it was taken (like bubonic plague) as a judgment upon the wicked; and since the spread of syphilis was linked to sexuality, this fit well within a Victorian value system. The tragedy is that the disease came to be seen as a moral issue, rather than as a medical one.

In the case of AIDS, the manner of prevention is well known (condoms), but so many taboos still pervade sexuality in our society that disseminating the information, or the prevention itself, is invariably a source of controversy. Again, like syphilis, by shifting the focus to sexuality rather than to the disease, we impede our own attempts to cope with it.

Of course with AIDS, the issue of sexuality is overlaid with homosexuality, since the earliest spread of the disease in America was among promiscuous male homosexuals. Thus, rather than being a strictly rational public health issue, AIDS has virtually become a public referendum on the acceptability of homosexuality. This is strikingly similar to the syphilis epidemic of decades ago, in which the public health issues became confused with the moral issues. Back then, even heterosexuality was considered dubiously moral; now, the focus of the moral issues has changed, but it still serves as a distraction from the real issue, which is medical.

Like syphilis, association of the AIDS dialogue with morality rather than with health has resulted in misinformation. Much of this is reflected in a concern over casual transmission of the disease, which, though conceivable, is so unlikely as to be a negligible factor in daily life. Like both bubonic plague and syphilis, we know that segregation and stigmatization of affected individuals won't check the spread of the disease, for it will simply drive affected individuals "underground." This will not only fail to end the epidemic, but will make risk factors and accurate data more difficult to obtain.[30]

AIDS has forced us to come to grips with fundamental issues in American life, for example, in the meaning, nature, and risks involved in sexuality. It also raises similar issues to those faced by earlier generations under different political systems: the role of the state in promoting, providing, and protecting public health; and the relative importance of individual rights and civil liberties as against public health and the common welfare.[31]

CULTURE AS TECHNOLOGICAL FIX

Just as many aspects of contemporary culture have promoted the spread of AIDS, and have impeded efforts to deal with it, we expect that culture, in the form of medical technology, will ultimately rid us of it. Though there may be cryptic biological factors affecting the susceptibility or immunity of individuals to the disease, ultimately our expectation is that the solution to this biological problem will emerge from the cultural realm. That is, after all, how we humans have solved our environmental challenges for even longer than we have been human.

Cultural solutions to problems, however, always come with a price tag: more problems for the next generation to solve, caused by the present generation's solution. Sometimes the problem may be as simple as complacency following the elimination of one problem—such as the optimism in the field of public health after the eradication of polio and before the outbreak of AIDS. Alternatively, it can be like hubris in the Greek tragedies: the evolution of resistance to antibiotics in pathogens, bringing the disease back, as in the new forms of gonorrhea and tuberculosis.[32]

More often, however, cultural solutions to health problems lead to other problems that are not so clearly medical as they are social. People in the industrialized nations are living longer than humans ever have, for example. The advantages are self-evident, but along with this gift of extended life comes the "aging of America." Health care costs rise, as the elderly require more extensive treatment for their illnesses; and where in more traditional societies the elderly have roles in raising their grandchildren and great-grandchildren, in America they generally do not live near their descendants, and are often alienated from productive roles in contemporary society. While this is widely appreciated to be a social problem, it is unclear what the answers are.

Imagine, however, the social consequences of extending the human life span to 120, as some scientists and science journalists occasionally predict with utopian enthusiasm.[33] For a society that has great difficulty integrating 80-year-olds, how attractive a proposition is a 120-year-old life span? This is not to say that scientific work should not be encouraged in that direction; only that at least as much effort should be directed at predicting, evaluating, and confronting the consequences of technologically tinkering with our biology. Technological bio-tinkering has far more wide-reaching effects than merely somatic, as novelists from Mary Shelley to Michael Crichton have pointed out.

And more to the point, what about the increase in human biomass that would come as an automatic consequence of extending the life span by 50 percent? Is that desirable? Or might it be wiser to direct our attention to improving the quality of life within its current span, thereby encouraging people to reproduce less, and ultimately *reducing* the earth's human biomass?

NOTES

1. Herskovits (1938), Shapiro (1939).
2. There have always been elements of elitism and hypocrisy in the population issue. Most Americans who would agree with the viewpoint that overpopulation is the greatest problem facing our planet nevertheless would reserve for themselves the right to propagate. It is always others who are reproducing profligately; the world is always big enough, it seems, for one more Harvard man.
3. Lee and DeVore (1968).
4. Konner and Worthman (1980).
5. Johnson and Earle (1987).
6. Whether the increase in population is a consequence or a cause of food production is a dispute of long-standing within anthropology, but the association between population increase and agriculture is clear: see Streuver (1971), Cohen (1977), Molleson, Jones, and Jones (1993).
7. Handwerker (1983).
8. Mauldin (1980), Coale (1983).
9. Westoff (1986), Coleman (1990).
10. Zinsser (1935), Crosby (1986).
11. Black (1992).
12. Cohen and Armelagos (1984), Cohen (1989).
13. Burkitt (1981).
14. Inhorn and Brown (1990), Gajdusek (1990), Krause (1992).
15. Gajdusek (1977), Alpers (1992).
16. Barker (1989).
17. McKeown (1988), Merbs (1992).
18. Zinsser (1935), Robins (1981), McNeill (1976), McEvedy (1988), Slack (1989).
19. Luzzatto and Battistuzzi (1985), Bank (1985).

20. Motulsky (1960). See Thomson (1983) and Bell, Todd, and McDevitt (1989) for a review of the associations between diseases (especially chronic) and alleles for the human major histocompatibility complex (HLA); Davies (1993) for diabetes and HLA; Markow et al. (1993) for balancing selection and HLA alleles; Mourant (1983) for the tenuous association between blood groups and diseases.

21. Dean (1971) for porphyria; McKusick, Eldridge, Hostetler, and Egeland (1964) for Ellis–van Creveld syndrome.

22. Myerowitz (1988), Navon and Proia (1989), Grebner and Tomczak (1991), Beutler (1992).

23. Myrianthopoulos and Aronson (1966, 1972), Myrianthopoulos, Naylor, and Aronson (1972), but see Chase and McKusick (1972), Fraikor (1977), Spyropoulos, Moens, Davidson, and Lowden (1981), Spyropoulos (1988).

24. Meindl (1987), Jorde and Lathrop (1988), Tsui and Buchwald (1991).

25. Roberts (1981), Kunitz (1993). Since the causes of death vary by age group, as more people live to older ages the causes of death become more skewed toward diseases of old age.

26. Ward (1983).

27. Many genetic diseases are far less deterministic in their onset and course than Tay-Sachs, sickle-cell, and cystic fibrosis. The variability of such diseases leads to a probabilistic assessment of risk, independently of knowledge about individual genotypes. Schull (1993) reviews the difficulties in inferring a genetic etiology from the observation of ethnic differences in prevalences of disease. Often the genetic link is conjectural, made in the absence of a mechanistic model and identifiable DNA sequence; and often the genetic cause is merely inferred from a correlation of ethnic difference and health difference. Regardless of the actual etiology of the disease, however, observed associations between ethnicity and disease are useful as "risk factors" in health care.

28. Piot et al. (1988), Curran et al. (1988), Merson (1993).

29. Rice (1929), p. 183.

30. Brandt (1988), Quetel (1986 [1990]).

31. Dickens (1988), Walters (1988).

32. Culliton (1976), Chase (1982), Bloom and Murray (1992), Ryan (1993), Rothenberg (1993).

33. For demographic changes in the age structure of modern populations, see Olshansky, Carnes, and Cassel (1993). On extending the human life span, see Leaf (1973), Barinaga (1991), Rusting (1992).

CHAPTER

12

Human Traits: Heritage or Habitus?

To understand human variation, the biology of humans must be dissected into those characteristics that are uniquely human and those that are shared with other creatures. Some universal traits, such as grasping and suckling in infants, are obviously part of our heritage. Often it is difficult to tell whether a behavior, such as infanticide, is indeed homologous to a similar behavior in nonhuman primates.

We finally come to the issue of longest standing in the human sciences: What is the relationship between patterns of genetic diversity and behavioral diversity in the human species? To approach this, we must distinguish between polymorphism and polytypism genetically, and between behavioral diversity *within* a social group (personal and idiosyncratic) and behavioral differences *between* social groups (cultural and historical).

There are certainly productive avenues of inquiry concerning the ways in which patterns of human behavior may correlate with ecological variables. Contemporary human behavioral ecology does not, however, postulate that differences between cultures are the result of differences in their gene pools. Rather, the diversity of human cultural forms results from specific historical processes operating on basically equivalent gene pools.

We can consequently differentiate between mainstream behavioral ecology and hereditarianism. There is certainly little doubt that genes influence behaviors to some extent, and that people vary polymorphically for those genes. "Nature and nurture" aren't the issues; the causes of within-group and between-group diversity are the issues. Unfortunately, in criticizing "the social sciences," sociobiology in the 1970s often failed to differentiate between the patterned behavioral variation of people representing groups from different times and/or places (i.e., anthropology), and why people in the *same* time and place do different things (e.g., psychology).

Relating biology and behavior in the human species is probably the

most value-laden scientific endeavor. Consequently it has been a battle-
ground for many ideological armies, all of whom claim to speak for science. That is probably unavoidable. What we will try and do here (and
in Chapter 13) is to approach the subject in the context of patterns of
genetic variation.

AESOP AND DARWIN

The paleontologist William King Gregory observed that to study any
species involved distinguishing between those characteristics which it
has inherited passively from its ancestors, and those which mark it as
different from them. Gregory called those remnant features due solely to
ancestry "heritage," and those that reflect the new adaptation of the particular species in question "habitus." More recent terminology would
call the former "plesiomorphies" and the latter "apomorphies" (Chapter
1).[1] This is not a distinction of "genetic" versus "learned," but rather of
biologically "ancient" versus biologically "new" for the species.

When we study the human species, one critical explanatory factor is
the novelty of the trait in question. Is it a part of our ape ancestry—like
grasping hands and rotating shoulders—or something we have acquired
only in the course of becoming human—like rigid ankles and bare skin?
The former implies passive inheritance of the trait in question, and
therefore no explanation is required for its retention in humans. The latter, on the other hand, implies a change in the recent biological history
of our species, which indeed calls for an explanation. If other animals
could live without the trait, why couldn't we? What caused us to
develop it?

Unfortunately, these kinds of questions are not amenable to scientific
analysis in the common sense of the term. Science analyzes regularities
in nature; but the search for reasons why something happened is the
study of a *singularity* of nature. Consequently the study of human historical biology is a science different in its foundations from others. And
by virtue of having political stakes in the scientific answers it produces,
it is a field far more introspective than other historical sciences.

A major concern of this field is the nature of explanation in human
historical biology. To creationists, there was no need to explain the way
things got to be as they are—they simply have always been that way.
But to an evolutionist, the present has been shaped by the past, and to
the extent that a particular human feature is not now as it once was, an
explanation for it is required.

The explanation is obviously that it evolved. But how? Did it confer
an advantage to our ancestors, permitting them to thrive at the expense

of other members of their population? If so, what was the nature of that advantage? Since the answers to these questions lie in the remote history of our species, were not videotaped, and cannot be replicated, they are often discursive in nature. That is, historical explanations tend to be more narrative in their structure. To derive the evolution of human characteristics from the world of biology, one needs to refer to other species. Evolutionary explanations are comparative in nature.

And yet, relating human traits to those of other species is an ancient literary form in Western civilization—indeed in the oral traditions of nearly all cultures. It precedes evolutionary theory, and is independent of it. This mode of association is symbolic, based on a metaphorical relationship between the human and the other species. The owl is wise, the fox crafty, the ant industrious; we associate the origins of this mode of comparison with the fabulist of the ancient world, Aesop.[2] It is an association by analogy and tells us, A human is like an owl in this key way.

This mode of association is literary and symbolic, and is largely independent of taxonomic affinity. Darwin's contribution, on the other hand, was to introduce another kind of comparison: to show that there is a *special* relationship shared by humans and relatively few other animal species. This is the relationship of recent common ancestry, resulting in a shared groundwork of biology. The relationship here is no longer analogical or metaphorical, but homological—that is, due to common descent. Humans and certain particular species are fundamentally similar, not by virtue of metaphor, but by virtue of sharing a fundamentally similar biology they inherit from a recent common ancestor.

The result is that one can no longer make biologically relevant comparisons of humans to random taxa, as the pre-Darwinians did. Instead, a spotlight is shined on a restricted group of animals—in our case, primates—as the animals most closely related to us, and therefore biologically most meaningful in terms of explaining our own behavior. In other words, the owl may be "wise," but a rhesus monkey is much smarter, because its brain is far more like a human's, because it is a close catarrhine relative of ours.[3]

Darwin's contribution thus places strong constraints on what a reasonably scientific comparison of human features with those of other animals should be. A significant comparison is one between two close relatives, for their structures are likely to be homologous. A comparison between distant relatives, who share little common biology, is not likely to turn up biologically meaningful similarities. The reason is simple: since the two species are distant relatives, any biological similarity between them is very likely to be superficial and artifactual.

Consider, for example, the following passage from a popular work on human evolution by an ornithologist:

Among our darker qualities, murder has now been documented in innumerable animal species, genocide in wolves and chimps, rape in ducks and orangutans, and organized warfare and slave raids in ants.[4]

Is this true? Is this what the Darwinian study of human behavior tells us—that the "wilding" of the Central Park Jogger, the 28-year-old investment banker who was gang-raped and beaten nearly to death by a large group of black and Hispanic teenagers for mischievous fun in 1989, happens to ducks? The voyage to America into slavery—do ants really experience anything like it?

[T]he sense of misery and suffocation was so terrible in the 'tween-decks—where the height sometimes was only eighteen inches, so that the unfortunate slaves could not turn round, were wedged immovably, in fact, and chained to the deck by the neck and legs—that the slaves not infrequently would go mad before dying or suffocating. In their frenzy some killed others in the hope of procuring more room to breathe. Men strangled those next to them, and women drove nails into each other's brains.[5]

Does evolution really tell us that the stories of lives cheapened and degraded by the sugar and cotton plantations of the New World happens to ants?

Of course not—quite the opposite, actually. The passage about the ducks and ants is a reflection of (presumably unintentional) non-Darwinian biology. The author's intent is clearly to suggest some form of biological equivalence between, for example, warfare in ants and warfare in humans. This would ostensibly imply that warfare is part of our nature—not necessarily good, but found in other species, and therefore biologically interpretable.

But what is the nature of that biological equivalence? How basically similar is the biology of an ant to that of a human? Obviously, not very similar. At the most fundamental biochemical levels of comparison, human and ant systems are similar to one another, but their bodies are so substantially different that it is difficult to find any similarity at all. So what can we make of the fact that warfare is present in ants and humans? Is it conceivable that over the hundreds of millions of years since the divergence of humans and ants, they have evolved different kinds of skeletons, different numbers of paired appendages, different kinds of sensory systems, and different kinds of life histories—*but retained the same warfare*?

Probably not. One can talk of "warfare" in ants, but it is on the same order as talking about "legs" in ants. Ants indeed have paired appendages which support the animal, but these are structurally entirely different from human legs (Figure 12.1). Their similarities are biologically highly superficial, in essence, a semantic trick of giving parts of both organisms the same word "legs," because they serve a similar function. Likewise, the wing of a fly and the wing of a bird have biologically nothing at all in common—the fact that they are both called "wings," reflecting the fact that both sets of structures flap and propel the bearer through the air, is biologically highly misleading, for they are anatomically, developmentally, and genetically unrelated.

What, then, do we make of the invocation of "warfare" in ants? Given that it is difficult to find *any* specific structures that are homologous in ants and humans, we are obliged to assume that warfare is not homologous either. What is here being labeled "warfare" in both species is an analogous rela-

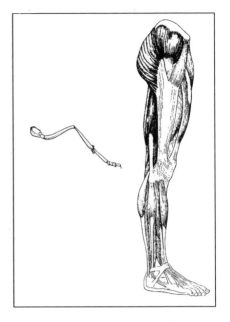

Figure 12.1. Leg of an ant and a human (not to scale).

tionship. Like "wings," it is merely wordplay—an identical label attached to structures that are superficially similar, and fundamentally entirely different.

What the quotation suggests, then, is a pre-Darwinian view of the evolution of human behavior, one in which taxa can be chosen irrespective of their biological relationships, deceptive similarities can be noted, and Darwin's own contribution can be ignored.

It serves another purpose as well—to overstate the affinities of human behavior to that of other animals. While the invocation of ants, ducks, and wolves appears to give the analysis of human behavior a very broad naturalistic base, the base narrows when we recognize that these are not biological similarities being noted at all, but *symbolic* similarities. They are basically linguistic artifacts, the consequence of putting the same label on different products. In other words, it downplays the uniqueness of habitus and implies that all of these human behaviors are due to heritage. Heritage may indeed play a role in understanding human behavior, but the biological arguments for it in a post-Darwinian world are obliged to come from our close relatives, the ones whose biology is similar to our own.

When we confine ourselves to our close relatives, however, we find (1) such a difference between human behavior and that of the apes, and (2) such a diversity of behaviors among the apes that it becomes difficult to argue for much of any human behavior being the result of heritage, rather than of habitus.

As an example, the pygmy chimpanzee or bonobo (*Pan paniscus*) commonly displays a suite of behaviors unknown among other primates. For one, two females may stimulate each other sexually (a behavior known as "genital-genital [or g-g] rubbing"). For another, a female may initiate sexual activity in de-escalating an aggressive encounter, or in acquiring food from another individual. The parallels to the range of human behavior are obvious. But what is the biological connection? After all, these two suites of behaviors are unknown in common chimpanzees and in gorillas.

One could infer they are homologous, and link pygmy chimpanzees and humans as closest relatives on the basis of sharing these behavioral features. But the closest relative of the pygmy chimpanzee is the common chimpanzee, not the human—an inference well supported by genetic evidence. So the initial impact of this behavioral information is not to attest to a special closeness of specifically pygmy chimpanzees and humans, but to attest to a considerable breadth in the behavioral repertoire of the genus *Pan*.

Again, however, we have to ask with reference to the genus *Homo*, "Are the behaviors we recognize of ourselves in pygmy chimps part of our own behavioral heritage or habitus?" We can apply phylogenetic reasoning to the problem. If it is heritage, then the behaviors must have been present in *Homo* and *Pan* and then subsequently lost in *Pan troglodytes*. Behavior is not directly preserved for us in the paleontological record. But there is a simple test of this hypothesis. If the trait were originally present in *Homo* and *Pan*, it stands to reason it would have been present in *Gorilla* as well, since these three genera originated at about the same time from a hominoid of the Miocene, about 8 million years ago. And yet the gorilla does not seem to have the behavior either.

Ultimately the hypothesis we are left with is that these behaviors have been recently acquired (habitus) in parallel in the human and pygmy chimpanzee. The significance of these kinds of behaviors remains unknown, but it seems unlikely that we learn much about either homoeroticism or about commoditizing sex from studying the behavioral repertoire of pygmy chimpanzees.

SEX AND THE SINGLE FRUITFLY

One of the pitfalls of cross-species comparative studies of behavior stems from our infatuation with understanding the roots of our own behavior. When a behavior is found that appears to have a parallel in humans, it is widely assumed that the other species represents a primitive, natural state—that we have not merely observed some unique and bizarre specialization of this species.

A 1948 paper on the mating habit of fruitflies shows up this problem quite well. A. J. Bateman sought to prove "why it is a general law that the male is eager for any female, without discrimination, whereas the female chooses the male."[6] Bateman elegantly demonstrated experimentally that for a male, the number of offspring rises linearly with the number of mates the male has; but in females it levels off after one mating. Bateman's explanation lay in the basic nature of male-female differences: sperm are cheap, eggs are expensive; therefore, it is in a male's reproductive interest to spread his seed widely, but not in a female's reproductive interest to do so. That this happened to reinforce some powerful sexual stereotypes in European society (female fidelity, male wanderlust) was less relevant than the fact that those stereotypes were demonstrable biologically in fruitflies. And obviously, the explanation was as applicable to humans as to fruitflies.[7]

The shortcoming lies in a throwaway phrase in Bateman's paper. Focusing on "the greater dependence of males for their fertility on frequency of insemination," Bateman goes on to note that "[t]hough this will clearly apply to all animals in which the female can store sperm, it can be shown that it is in fact an almost universal attribute of sexual reproduction."[8] Perhaps so, but the specific issue is how broadly applicable these experimental results from fruitflies may be. After all, an important aspect of fruitfly reproduction is that females have an organ known as a spermatheca, which stores sperm—so that they do not need to mate again after the first time to have their eggs fertilized.

Clearly, then, the female fruitfly has specializations that make it particularly unnecessary to seek multiple copulations to maximize its reproductive output. In other words, the fruitfly results are strongly bound up in fruitfly habitus—uniquely derived aspects of its biology that have no homologue in humans. How can this legitimately be extended to humans when the biological system under examination involves fruitfly specializations, not generalizations?

The short answer is, it can't.

The relationship of male to female variance in reproductive success is simply not sufficiently comparable biologically between flies and people as to render such an extension meaningful. Certainly there is a *potential* difference, with males *in theory* being able to have far more or children than women. In practice, however, this requires empirical demonstration demographically, and probably only actually holds in a small number of social and historical situations.

What, then, can we say about human sexuality as a result of this fertility imbalance between male and female fruitflies? Splendidly little. The fruitfly data appear somewhat seductive because they coincide with Western views of the sexual double standard: men as philanderers, women as homebodies. But the biological connection to humans is a very weak one, for it is a study of fruitfly specializations. It is certainly plausible to think humans are specialized in the opposite direction. In humans, the cultural institution of marriage, in its myriad forms, helps keep the actual relative fertilities of males and females quite similar, in most societies and most families.

RAPE AS HERITAGE OR HABITUS

We saw in an earlier quotation an attempt to "naturalize" murder, genocide, rape, and slave raids by finding analogs in the animal kingdom. As we already seen, the connection between the human behavior and the animal behavior is non-Darwinian, and is made in spite of the absence of shared basic biology.

Where human behavior is concerned, history tells us that explanations for it imply social and political agendas, and biological explanations for human behavior have implied them especially well. Whether they are promoted consciously by scientists or out of simple naivete is impossible to say in any specific case. Nevertheless, the suggestion that a criminal was "just doing what comes naturally" certainly serves the purpose of trivializing the crime. If indeed the crime was the result of a natural, biological propensity, it may be judged wrong, but perhaps not quite so bad as if it were judged to be an "unnatural" act. With such social values at stake, a scientist writing about the "scientific" basis of human behavior is denied the luxury of moral or ethical neutrality. A strong burden of responsibility, history shows us, must be borne by those claiming to speak about human nature in the name of science. It is not like a scientific pronouncement on clam or bird behavior; it affects people's lives.

We read in a popular current textbook of ethology about the scientific, evolutionary explanation of rape:

> [S]exual selection in the past favored males with the capacity to commit rape *under some conditions* as a means of fertilizing eggs and leaving descendants. According to this view, rape in humans is analogous to forced copulations in *Panorpa* scorpionflies, in which males excluded from more productive avenues of reproductive competition engage in a low-gain, high-risk alternative. Male *Panorpa* that are able to offer material benefits to females do so in return for copulations; males that cannot offer nuptial gifts attempt to force females to copulate with them. Human males

unable to attract willing sexual partners might also rape as a reproductive option of the last resort.[9]

How are humans like scorpionflies? In rather few ways, but apparently in one biologically significant way: males force themselves on females sexually, in order to pass on their genes. Again we are presented with the suggestion that this human behavior is part of our biological heritage, reflected in the similarity to scorpionflies, and not something peculiarly unique to the human species, our habitus.

The validity of this scientific explanation rests ultimately with how comprehensive and reasonable it is. The first aspect to note is that the explanation assumes a single over-riding purpose for sexual activity: reproduction. Here we find an interesting concordance between this ostensibly scientific view of human sexuality and the puritanical views of some conservative theologies: sex is "for" reproduction. This has the implication that all non-reproductive sexuality is not only immoral, but "unnatural" as well—including homosexuality; oral and manual stimulation; sex during menstruation and pregnancy and after menopause; and sex while using birth control methods. In other words, it would render "unnatural" the great bulk of sexuality in the human species, which is *ipso facto* actually "natural" for the species.

What is the alternative? That human sexuality far transcends reproduction, as we have already seen. Human sexuality is about many other things in addition to reproduction—other things of such importance that the reproductive nature of sexuality can easily be considered a trivial component of sexuality. This is, again, an important part of our habitus.

To return to the scorpionflies, then, male and female scorpionflies have sex in order to perpetuate their genes, and sometimes a male forces himself on a female as a last-ditch effort to pass his genes on. In humans, though, if reproduction is only a small part of sexuality, then to assume all sexual acts are reproductive in nature is rather poor scientific reasoning. Again, it has the seductive benefit of harmonizing with cultural stereotypes, and that is what makes the assertion in the name of science so pernicious.

One need only skim a daily newspaper to find that the words most closely associated with "rape" are not "love" or "sex" or "baby," but "murder" and "torture"—as in "rape and murder" or "beaten, tortured, and raped." One will also find that a significant proportion of rape involves men and men (such as in jails), or men and children. Obviously the chances of reproducing here are negligible for the rapist. An explanation that assumes rape is about reproducing cannot begin to account for these acts—for this "scientific" reasoning only purports to account for specifically heterosexual rape involving post-pubertal vaginal intercourse. All other kinds of sexual violations apparently can be safely ignored by this "scientific" explanation. Indeed, so can rape by married men; it is specifically and exclusively here a singles crime as well.

What is the alternative explanation? That rape is not exclusively, or even principally, about reproduction—it is about power. It is fundamentally about the physical domination of a conspecific; and the fact that some of it involves men and women and vaginal intercourse is largely incidental. Intercourse may be simply the means to effect or assert domination over the victim. This would make rape largely independent of the marital status of the rapist and sex of the victim, and would make the prospect of reproducing a triviality to the rapist. The rape-as-reproduction hypothesis takes rape as a part of our heritage; the rape-as-power hypothesis takes it as habitus. The fact that it is largely unknown in our close relatives (with the exception of the orang-utan) makes it more likely that rape is a specifically human characteristic.

Again, scratching slightly below the surface of what are presumptively value-neutral scientific judgments, we find a world of culturally loaded assumptions. While it is no shame to put forward an inaccurate hypothesis, it may give us pause when such hypotheses are proposed to be value-neutral and more "scientific" than alternatives.

PROXIMATE AND ULTIMATE CAUSE IN BIOLOGY

One justification given for the rape-as-reproduction hypothesis is that it is explanatory as an "ultimate cause" in human behavior. The distinction between proximate and ultimate cause is one of the most widely misunderstood philosophical concepts in this science, unfortunately.[10]

A student of human behavior does not want to know why humans behave. Humans behave because they are multicellular animals, and can't photosynthesize like plants. They are consequently obliged to move around and do things. The question of interest is instead: Why do humans do some things and not others? In other words, what we want to explain is *variation* in human behavior.

Why do humans rape?, therefore, is really the question What motivates some humans and not others to rape? To answer that question with a pan-human universal—the desire to procreate—is to explain a variable with a constant. It has little explanatory power: in explaining both the phenomenon A and the phenomenon not-A, the explanation by recourse to a constant of human biology tells us nothing about the differences between those two states. If they have the same cause, how does one become A and one become not-A? That is the interesting scientific question: the search for a proximate cause of the behavioral differences.

Science is, in general, the analysis of the material causes of things, the study of processes and mechanisms in nature. Our conception of it is largely the result of the revolution wrought by Isaac Newton. Prior to that revolution the hand of God was seen in the daily operations of the physical universe, and subsequent to it those daily operations became subject to entirely material forces, while the hand of God was removed to the simple enactment of those forces. The Newtonian revolution, in effect, transformed God from a proximate cause of physical phenomena to an ultimate cause. In other words, physical laws keep the planets in their orbits and make apples fall to earth in a particular way; God creates the physical laws and then steps out of the way.

An ultimate cause is the cause of the cause, and it is scientifically problematic. First, discussion of ultimate causes leads to an infinite regression: an interest in the cause of the cause leads to an interest in the cause of the cause's cause. Ultimately, then, all phenomena in the physical universe are explained by the Big Bang, which seems hardly adequate to those with an interest in human behavior. Second, ultimate causes tend to end up in mysticism. What caused the Big Bang? If everything that causes something in the universe is itself the effect of another cause, then, reasoned Aristotle, God is the logical beginning of that chain: the uncaused cause, the unmoved mover.

If this sounds a little cosmic, that is why contemporary science generally restricts itself to the study of proximate causes, of mechanisms and processes. While

> Nature is but a name for an effect,
> Whose cause is God.
>
> —William Cowper

speculation on the nature of ultimate causes (i.e., God) was still considered scientific in Newton's day, the major advance was in teasing apart the scientific analysis of proximate cause from speculation on ultimate causes.[11]

Darwin, indeed, crafted his idea of natural selection in accord with this model. Prior to Darwin, species were considered to have been molded by the hand of God with their peculiar specializations and adaptations. Darwin conceived of natural selection as an efficient, proximate cause of the differences between species. This, on the Newtonian model, gave biology a material scientific basis. But what started things off? That was not a question that Darwin would deal with directly, for it was a question of ultimate cause. The concluding sentences of *The Origin of Species* make it clear that he saw a role for ultimate causes, and that they lay in another sphere:

Thus, from the war of nature, from famine and death, the most exalted object which we are capable of conceiving, namely, the production of higher animals, directly follows. There is grandeur in the view of life, with its several powers having been originally breathed [by the Creator] into a few forms or into one; and that, whilst this planet has gone cycling on according to the fixed law of gravity, from so simple a beginning endless forms most beautiful and most wonderful have been, and are being, evolved.[12]

The bracketed phrase "by the Creator" was added by Darwin in the second edition to make his meaning even less ambiguous. Natural selection is a *proximate* cause of differences between animal populations; the cause of competition, life, and natural laws is God. And God is not the subject of this investigation; our subject is the diversity of life.

Post-Darwinian scholars have limited themselves exclusively to the study of proximate cause. Following Darwin, this means that adaptive differences between populations are the result of the action of natural selection.

Oddly, the justification sometimes given for studying rape as a reproductive strategy is that reproduction is an ultimate cause, not a proximate cause of the phenomenon. In this sense, however, the proffered "ultimate cause" is simply an alternative explanation for the phenomenon, not an ultimate cause in the ordinary usage. Indeed to invoke selection as an ultimate cause is to use it in a manner precisely the opposite of what Darwin intended! Natural selection through differential reproduction is a proximate cause of the adaptive biological differences between populations; its analog, sexual selection, is a proximate cause of the non-adaptive (in the sense of not tracking the environment) biological differences between sexes.

There is, of course, a sense in which genes are an ultimate cause of human behavior. If we had the genes of a cow, we would behave differently. We would behave like cows. Human genes compel us to behave like humans. But human behavior is extremely diverse, and it is that diversity we wish to explain. Why do we eat? is certainly explicable by recourse to our genes, but it is an inane question with a trivial answer. Why do some people eat rare filet mignons and other people don't? still has an ultimate trivial cause in the genes, but the answer—the proximate cause— lies elsewhere. It lies in economics, socio-cultural history, and personal experience.

Likewise with rape, which has the confusing aspect of a sexual component, which we associate with reproduction. Nevertheless, genetics is not likely a proximate cause of rape, for rapists are an exceedingly heterogeneous lot; as an ultimate cause, it fails to tell us anything interesting about specific instances of the phenomenon. What little it tells us of the phenomenon in general is not amenable to scientific test; so while it may not be obviously false, the "rape-as-reproduction" hypothesis is framed in such a way that truth and falsity have no meaning. The issue is not so much to deny that rape has some sexual component, or that sexuality is in some sense reproductive, but rather to deny that this constitutes an adequate scientific explanation of the phenomenon, or in more extravagant formulations, *the* scientific explanation. Like eugenics of the 1920s, it looks and sounds scientific, but is simply a set of cultural values spliced into a theory of evolutionary biology.

THE ASPHALT JUNGLE

Many of these issues combined in an illuminating episode in 1992. Frederick J. Goodwin was head of the Alcohol, Drug Abuse, and Mental Health Administration, and a strong supporter of research on the behavior of nonhuman primates. He was promoting the study of violence from an evolutionary, scientific perspective, and gave a speech citing the high rate of mortality for male rhesus macaques in support of it:

> Now, one could say that if some loss of social structure in this society, and particularly within the high impact inner-city areas, has removed some of the civilizing evolutionary things that we have built up ... that maybe it isn't just the careless use of the word when people call certain areas of certain cities jungles.[13]

In the rhubarb that ensued, Goodwin lost his job under diffuse accusations of racism.[14] Whether there is actually racism or genetic determinism here is unclear, but what is clear is that the speaker was attempting to invoke a scientific, evolutionary explanation for the phenomenon in question, and there are four fallacies evident in the argument.

First, it assumes a unilinear evolutionary mode, a single trajectory leading from rhesus macaques to government bureaucrats, and from which inner-city youth have slipped backwards a few notches. The key word here is "loss"—the inner-city youth have regressed, as it were, into a primate vortex. The assumption is that we are witnessing the *loss* of civilization, rather than a *reaction to* civilization; or rather, in many cases, a reaction to the back of civilization's hand.

Second, there is an expression of elitism here: One never encounters such "evolutionary" explanations for white-collar crime. Whether the Watergate burglars were more like a group of ring-tailed lemurs than *you* are is not apparently of interest to anyone, and has never been raised. Some humans are greedy—for cultural, symbolic things like money or power—and they use the means at their disposal to satiate that greed. But it's the *lower-class* manifestations that are likened to the nonhuman primates, not the upper-class manifestations.

> Successful and fortunate crime is called virtue.
>
> —Seneca

When Charles Davenport defined "feeble-mindedness" as the allele responsible for crime, it was not only specifically for lower-class crime, but it was an atavistic mutation as well:

> The acts of taking and keeping loose articles, of tearing away obstructions to get at something desired, of picking valuables out of holes and pockets, of assaulting a neighbor who has something desirable or who has caused pain or who is in the way, of deserting family and other relatives, of promiscuous sexual relations—these are crimes for a twentieth-century citizen but they are the normal acts of our remote, ape-like ancestors and (excepting the last) they are so common with infants that we laugh when they do such things. In a word the traits of the feeble-minded and the criminalistic are normal traits for infants and for an earlier stage in man's evolution.[15]

Third, Goodwin has confused in his invocation of "civilizing evolutionary things that we built up" the evolution of culture and of species. After all, the kinds of things that distinguish civilized Washington bureaucrats from uncivilized Puerto Rican gang members are categorically different from the kinds of things that distinguish Puerto Ricans—civilized or uncivilized—from rhesus monkeys. The former are the products of *social* history, the latter, of *biological* history. The processes of evolution are different, though the words are the same—again we are confused by the application of the same word to two different classes of phenomena.

And finally, why the macaques? Certainly we have been phylogenetically distinct lines for the last 25 million years or so. They are among our close relatives, but not nearly as close as the gibbons, practitioners of traditional family values, or the gorillas, who were vegetarians long before it became popular among yuppies. The byte of macaque demography can't transcend the tens of millions of years of evolutionary divergence between our species and theirs: Goodwin was comparing different habitus, not common heritage.

HUMAN BEHAVIOR AS HERITAGE

William King Gregory defined the habitus of a species as "all those characters which have been evolved in adaptation to their latest habits and environments."[16] Our most distinctive habitus is the generation of culture (Chapter 2), which through its social and symbolic nature molds and transforms our lives. Most human behavior is cultural, and culture is the habitus of our species.[17]

Most of our species' behavior occurs in the complex cultural universe of status, power, self-identification, education, economics, ambition, imagination, and love. Most of our behavior, therefore, is studiable only as habitus; we can demonstrate little in the way of biological homology with other species. An aggressive encounter between two chimpanzees has little in common with an encounter between a Nazi brownshirt and a Jew in 1936 Germany or a gay-basher and his victim in 1983 America. The latter encounters are about group behavioral differences, and are charged with emotional symbolic power, and may be between two individuals who have never met before and want nothing from one another. The chimpanzee encounter is about the immediate circumstances, and is between those two individuals.

On the other hand, some human behaviors are arguably heritage, which Gregory defined as "all those characters which were evolved in adaptation to earlier habits and environments and which were transmitted in a more or less unchanged condition, in spite of later changes in habits and environments." Smiling, sobbing, the grasp of an infant (presumably to cling to its mother's fur though she lacks it), the physiological changes associated with tension are all presumably homologous to similar behaviors in our close relatives.

Most behaviors, of course, fall into a gray zone, in which it is unclear whether they are homologous between human and ape, and thereby part of our heritage; or nonhomologous, and part of our habitus. One interesting example is infanticide, known in primates and in humans. When a male langur monkey invades a new group he sometimes kills the babies in the group, and impregnates the females; the behavior appears to be directly explicable as a reproductive strategy on the part of the male.[18] Infanticide also occurs in chimpanzees, but not associated with male takeovers; it seems to be a by-product of aggression against the mothers, and often involves cannibalism of the dead infant.[19] And in humans, the best-documented cases of infanticide are committed by the mother herself, reflecting decisions based on economics or social pressure.[20]

Are these all homologous behaviors, part of a human heritage? Or are they, like wings of insects and birds, fundamentally different structures to which we can merely attach the same label? Ultimately, there is no litmus test of homology, but we have to be impressed by the contextual differences and by the differences in proximate cause in these species.

At root, then, the application of evolutionary biology is critical for an understanding of human behavior. However, applying evolutionary biology to human behavior clearly does not imply that particular human behaviors are strictly comparable across closely related taxa, much less across distantly related taxa. Though occasionally such phylogenetically long-distance comparisons seem biological, and consequently scientific, they are actually non-evolutionary and pre-Darwinian in nature. The application of evolutionary biology involves inferring homology, and determining whether the structures being compared are indeed biologically fundamentally similar or not. If not, then giving them the same name is linguistically useful, for the structures are associated by analogy—but biologically confusing. And in the study of human behavior, the necessity is great for reducing confusion, because the quality of people's lives may depend on the nature of the "scientific" explanation, as political decisions come to be made, based those explanations.

NOTES

1. Gregory (1913, 1951). For contemporary cladistic terminology, see Eldredge (1982, 1985), Wiley (1981).
2. Simon (1978), Kitcher (1985:13).
3. There is an obvious anthropocentrism in calling the rhesus monkey smarter than an owl because its brain is like ours. We are deciding that mental processes approximating our own constitute smartness, for we are the smartest species by the criteria we apply. As Julian Huxley (1960:59) exclaimed on being accused of anthropocentrism by astronomer Harlow Shapley, "I should hope so! After all, we are *anthropos*."
4. Diamond (1992:170).
5. Bennett ([1966] 1962: 40–41).
6. Bateman (1948:352).
7. Trivers (1972), Daly and Wilson (1983). See Hrdy (1986) for a critique of the direct extrapolation of Bateman's work to humans.
8. Bateman (1948:364).
9. Alcock (1993:554), emphasis in original. This follows the contentious suggestion of Thornhill and Thornhill (1983, 1992).
10. See Barash (1993) for an example of this confusion. While noting that the two modes of explanation are not mutually exclusive, Barash advocates an "ultimate" approach that fails to explain the presence/absence of the phenomenon. See also Beatty (1994), Mayr (1994).
11. Reichenbach (1951), Hall (1954), Scriven (1959), Mayr (1961), Simpson (1963), Gilson ([1971] 1984), Rosenberg (1985).
12. Darwin (1859:490).
13. Quoted by Hilts (1992).
14. Goodwin subsequently became head of the National Institute of Mental Health.
15. Davenport (1911), p. 262.
16. Gregory (1913), p. 268.

17. Though anthropologists use *culture* to denote a symbolically mediated, historical system of ideas possessed by a society, the term is occasionally misapplied to those behaviors learned by an individual (e.g., Cavalli-Sforza and Feldman 1981; Bonner 1982). This would obviously extend it beyond the human species. Though culture is learned, however, that is only one property of it, not an adequate definition of it. Culture, as used by anthropologists, transcends the knowledge of individual people: English is learned, but no single person knows it all; constructing computers is part of our culture, but no single individual can build one from raw materials. It is consequently useful to distinguish between culture and learned behavior, the former being a special case of the latter.

18. Hrdy (1977).

19. Goodall (1986).

20. Hausfater and Hrdy (1986).

CHAPTER

13

Genetics and the Evolution of Human Behavior

Human behavior, like all phenotypes, has a genetic component, though it is diffi-
cult, if not impossible, to match genes to behaviors. Genetic differences among
humans may explain part of the range of behavior of people in any group. Vari-
ation in behavior between natural human populations has no detectable genetic
basis; between groups of people defined in other ways, it is difficult to tell, but
people want an easy answer, which comes at a social cost. The search for gener-
alizations about human nature is undermined by human biology, the prevalence
of genetic polymorphism. Speculations on human nature have tended to be
derived from the false premise that normal behavior can be narrowly delineated.
The essence of human nature, however, is to be multifarious.

ON THE NUMBER OF MICHAEL JORDANS
IN THE KNOWN UNIVERSE

A noted anthropologist with a reputation for iconoclasm was quoted
in the journal *Science* on the differences between races: "There is no
white Michael Jordan, one of the greatest basketball players ever to play
the game, nor has there ever been one."[1]

The statement is obviously true. The only Michael Jordan that has
ever been known to exist has a significant component of equatorial
African ancestry, which defines him as "black," rather than "white."
The truth of this statement lies, however, in its triviality. To the extent
that it is a statement about the uniqueness of Michael Jordan, its truth
is indisputable; there is no white Michael Jordan. Yet, until quite
recently there was no black Michael Jordan either: Not only is he genetically unique (as are we all), but he is a singularly exceptional athlete;
and comparing a singularity to the absence of a singularity is not a scientifically useful comparison.

But that isn't really what the statement was about. It was intended to
mean something more: presumably, that blacks on the whole are naturally better basketball players than whites. To derive that implication,
however, Michael Jordan shifts from being an *extraordinary* black person

to a being a *representative* black person. Yet that statement is now strongly reminiscent of Count Arthur de Gobineau's observation that he knew of no Charlemagne, Caesar, or Galen among the tribes of Native Americans (Chapter 4). And it is absurdly easy to falsify: for Charlemagne, Caesar, and Galen were not exactly typical representatives of their groups. Further, on what basis can we say with any confidence at all that another Galen was not born to a 14th-century Huron woman, or another Michael Jordan to an 8th-century Danish woman? We have no records to check. And we have no way of knowing how those children could even have been identified. In the absence of a Greek philosophical/scientific tradition or the 20th-century game of basketball, such prodigies would go unnoticed, their skills uncultivated, their potentials untapped.

Greatness, after all, is very eclectic in its nature. Consequently, greatness is strongly defined by opportunity and by culture, as H. L. Mencken argued about Babe Ruth.[2] Without baseball, it is hard to imagine Ruth's prominence, especially as an athlete, given a physique rather un-athletic by most standards. And yet, Ruth was both unique and great, and there was no black Babe Ruth—until Hank Aaron.

So what does this imply about the lack of Michael Jordans among people of predominantly or exclusively European ancestry? Exceedingly little. One is here comparing the presence of a performance with the absence of a performance, and inferring the presence of ability and the absence of ability. As we noted in Chapter 6, that is a false deduction. Performance implies ability, but lack of performance need not imply lack of ability, for many factors comprise performance. While all of the Michael Jordans have been black, all of the Bob Cousys and Pete Maraviches have been white. All of the Muhammad Alis have been black, and all the Rocky Marcianos white; all of the Ella Fitzgeralds have been black, and all the Barbra Streisands white; all the Bob Gibsons black, and all the Sandy Koufaxes white; all the Paul Robesons black, and all the Laurence Oliviers white. But comparing the best, or most extreme, members of a group to one another says nothing about the nature of the groups themselves.

COMPARING GROUPS OF PEOPLE

Generally, there are three comparisons obtainable when populations are compared (Figure 13.1). In the first place, for some ordinary variable of phenotype, one may find extensive overlap between the distributions of the variable in the two populations, with a small difference in the average value. This is often, for example, what we find for height or IQ, or the skin color of two populations 1000 miles apart. In this situation, most individuals fall within the area of overlap between the two populations' distributions; and virtually any individual can be drawn from either population. The cause of the mean difference is usually open to dispute, but the difference itself is a statistical characteristic of the populations.

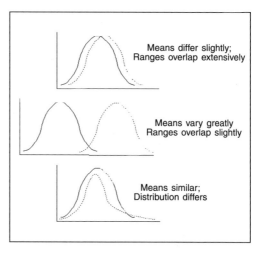

Figure 13.1.   Population comparisons.

Second, the populations could differ almost wholly in their distribution, with the means being very different. Here, any individual is easily assignable to one or the other population. An example might be the height of the pygmies of Zaire, compared to their neighbors, or the skin color of two populations 5000 miles apart.

Third, the averages may be roughly similar, but the distribution of variation may be different, with one group having more extreme members than the other. In their primitively non-social conceptions of culture change, some of the more liberal-minded of the eugenicists conceived of intelligence as distributed this way. They reasoned that innovations changed cultures, that geniuses were innovators, and that consequently the most technologically advanced societies must have been composed of the highest proportions of geniuses. People from the more "backward" lands might possibly be, on the average, as smart as a western European, but, lacking the most extremely smart people, those populations ended up as cultural backwaters. This, of course, is undermined by considering the real processes of social history in the analysis of culture change.

Most of the differences detectable between human populations tend to be of the first type, with much overlap, and a small difference between the two averages of the populations calling for an explanation. Unfortunately, when most people hear that two populations are different, they tend to hear the second type of situation. And the third type of situation really isn't about population differences at all, but rather about comparing the very few extremely deviant individuals at the ends of the distribution, the Michael Jordans.

WHERE ARE THE GREAT JEWISH BOXERS?

Every great sumo wrestler that has ever lived has been Asian. Does it follow that if we took a random African and a random Asian baby, and raised them identically, the latter would have the greater natural ability to develop into a prominent sumo wrestler?

Clearly the observation of achievement (exceptional performance as a sumo wrestler) is an inadequate basis for judging "racial" potential. Prominence at sumo wrestling is strongly contingent upon social and historical matters, which again highlights the asymmetry of evidence in inferring potentials from empirical observations. The fact that there have been great Asian sumo wrestlers means that Asians have the potential to be great sumo wrestlers, but the fact that there have *not* been great African sumo wrestlers does *not* mean Africans lack the potential for it. It may only mean they've never tried it.

Performance has many causes, only one of which is ability. Lack of performance does not necessarily imply lack of ability. Yet we seem to be far more interested in detecting ability—a highly metaphysical concept—than in accepting performance as its own standard.

Are blacks better at sports than whites? Certainly, to judge by contemporary performance. Yet we only see a preponderance of black athletes in certain sports: basketball, football, boxing, and baseball. We don't see it in golf, swimming, bowling, or tennis. Nor in sumo wrestling. Can we infer a difference in "ability" or "innate potential" here?

There is an important difference between those two categories of sports, namely that you can learn to fight, and thereby develop those aptitudes, virtually anywhere in America. But you can only learn swimming or golf and develop *those* aptitudes if you have access to a pool or country club. In Brazil, interestingly, where there is less segregation on the basis of degree of African or European ancestry, there are many competitive black swimmers, who do as well as their white competition.[3] We encounter prominent Latinos in baseball and boxing, but not in football or basketball.

Most likely, the prominence of black athletes has much to do with the ease of access into the middle class, and the kinds of options that are open to them. The prominence of major league shortstops from the Dominican Republic probably reflects more of how kids grow up there than about the distribution of native abilities in our species.[4] The number of Jews in major league baseball has declined precipitously in recent decades; there are Jews in the basketball Hall of Fame, but none of prominence now; one can find Jewish boxers of prominence from the

1920s, but today the phrase "Jewish boxer" sounds oxymoronic.[5] Have Jews gotten worse at sports?

The decline of prominent Jews in sports more likely is a reflection of other doors being opened than of that door closing. Where professional boxing is a major avenue of entry into the middle class, people are inclined to take it; but where medical school, law school, and other professions are open (as they were not often to Jews in the 1920s), fewer people naturally gravitate to boxing.

The case is unproven, and unprovable, since it is fundamentally a question of metaphysical genetics. How can we judge "natural ability"? Extrinsic factors dictate levels of performance, and we cannot measure intrinsic differences in ability. This is precisely the dilemma faced by IQ testers in the early part of this century (Chapter 6). The natural aptitudes required for baseball are coordination, eyesight, strength, and speed. To the extent that genetic variation exists for these things, there is probably far more polymorphism than polytypism, as is the case in the genetic systems that have been studied in the human species. Consequently, there is probably far more variation within groups than between groups; and to the extent that there are many aptitudes, caused by many genes, it is unlikely that their overall distribution is exceedingly skewed in one direction or another.

Thus, when a comparison is made between the best performers in sports to support the contention of superior abilities of their groups, we are inevitably left with a mass of contradictions. Since the game of baseball has many facets, there can be no fair single scale on which we can rank linearly Babe Ruth, Hank Aaron, Hank Greenberg, and Sadaharu Oh comprehensively to determine who the overall "best" was. Can we generalize fairly from the female basketball player, Nancy Lieberman, and male swimmer, Mark Spitz, to the innate superiority of a Jewish ancestry for athletes? Or from the long-distance runner, Grete Waitz, to the innate superiority of Norwegian women as marathoners?

Presumably not, for historical and social factors, as well as personal attributes, intervene to determine individual performance. Therefore, as regards blacks in professional basketball, one is obliged to retreat from "the Michael Jordan argument"—generalizing from the achievements of an outstanding individual—and look at the obvious numerical superiority of people of African ancestry in professional basketball. Might this now imply the superior ability of blacks?

Again, we have the problem that although this sounds like a scientific question, it is not. We can't measure ability; we can only measure performance. And since many things beside ability go into performance, there are too many variables to control in order to infer ability from per-

[ᵒ]ᵒᵣᵐᵃⁿᶜᵉ reliably. The numerical predominance of good professional black basketball players (as opposed to good white professional basket-ball players) may be a consequence of the fact that professional sports, and especially basketball, constitute an exceedingly risky way to earn a living. Though sports are an avocation for many (and a vocation for some) whites, it is nearly impossible to earn a living at it; and given other opportunities, it is a poor prospect for a livelihood. Not given other opportunities, however, it is probably as good a thing to train for as anything else. As Arthur Ashe notes:

> [Even the drug-implicated death of Len Bias] failed to dampen the enthu-siasm of tens of thousands of black youngsters who still aspire to profes-sional basketball careers at the expense of other, more viable, options. . . .
> The sport became an obsession in many black communities in the late sixties and early seventies. And why not? Basketball players were the highest paid sport athletes, and basketball courts were within walk-ing distance of nearly every black American.[6]

It is certainly not the case that natural aptitudes emerge without train-ing in black basketball players. In the case of Oscar Robertson, who led the NBA in assists for six seasons (supplanting Bob Cousy), and is the fifth-highest all-time scorer (just above John Havlicek):

> Robertson's motivation was no different than any other black youngster. "I practiced all the time . . . we didn't have any money and sports was the only outlet we had."[7]

Clearly, the contribution of "environment"—in this case, largely self-esteem and practice—is impossible to sort from those elusive, if widely invoked, natural aptitudes:

> [B]y the mid-1960s, the big black men brought their own distinctive style of play to the hardwood. While most of them had come from solid college experiences, they had learned to play in black environments where they impressed one another with the latest moves.[8]

Are blacks physically superior to whites, due to the construction of their bodies, in track events? Arthur Ashe called this a "ridiculous notion that was thought to be just the reverse before Eddie Tolan and Ralph Met-calfe won the sprints in the 1932 Olympics":[9]

> The fact that so many blacks are sprint record holders does not mean that blacks are better natural sprinters; but that more athletically inclined

blacks took an active interest in sprinting than did the general white population.[10]

All professional athletes have natural talents. It is these talents that are cultivated to enable an individual to earn a living as a professional athlete. If the issue is whether a superior natural talent exists in people of equatorial African ancestry for, say, basketball, there does not seem to be a reasonable way to generate an answer. The arguments in favor of it are specious; but ultimately the question as posed is simply another expression of folk heredity, the metaphysical genetics of natural aptitudes and uncontrolled experiments.

HOW DO WE ESTABLISH THE GENETIC BASIS OF A BEHAVIOR?

We observe that people are different. Some become ax murderers, and some become professors at Harvard. Could the Harvard professor have developed into an ax murderer? Or was there something constitutionally different about them, which all but fated one for one profession, and one for the other?

As phrased, the question is tricky, since it involves ostensibly comparing two specific life histories, and is therefore not terribly conducive to scientific analysis. Suppose we ask the question more broadly: Is there a constitutional difference by which to sort out a population of ax murderers from the Harvard faculty?

The question of whether there is a genetic basis to criminality has a long history in the social sciences. And for good reason: A straight answer immediately implies social policies, as was apparent to the eugenicists. The first and most obvious place to look is in the heads of criminals, for that contains the thoughts that lead to the criminal act. Unfortunately, it is difficult to look in the heads of criminals, so the next best thing is to look *at* the heads of criminals. A 19th-century student of the problem, Cesare Lombroso, maintained that there was a distinctly criminal appearance, which was ape-like, a throwback to savage, prehuman times. But applying statistics to the problem slightly later, the English Charles Goring found no significant physical differences between criminals and Cambridge students.

The Harvard physical anthropologist Earnest Hooton devoted the major work of his life to trying to correlate looks to criminal behavior, expecting that society would be aided ultimately in the prediction of criminality. In his monograph, *The American Criminal, Volume I*, Hooton believed he had found significant physical differences between criminals and a law-abiding group carefully chosen to control for race and geography. Hooton managed, unsurprisingly, to confirm all his assumptions:

> that it is from the physically inferior element of the population that native born criminals of native parentage [i.e., Anglo-Saxon] are mainly derived. My present hypothesis is that physical inferiority is of principally hereditary origin; that these hereditary inferiors naturally gravitate into unfavorable environmental conditions; and that the worst or weakest of them yield to social stresses which force them into criminal behavior.[11]

And the relation to eugenics is unhidden:

> An accurate description of an average gangster will not help catch a Dillinger. Certain theoretical conclusions are, however, of no little importance. Criminals are organically inferior. Crime is the resultant of the impact of environment upon low grade human organisms. It follows that the elimination of crime can be effected only by the extirpation of the physically, mentally, and morally unfit, or by their complete segregation in a socially aseptic environment.[12]

Kill them or put them away: those are the alternatives, within this framework, to the problem of crime. It is in their constitutional makeup, and is physically detectable; those can be our keys to dealing with those fated for the life of crime.

Hooton's second volume on the subject was never published. Reviewers pointed out that crime has two components: an objective behavior, and a subjective reaction to it, making it a criminal act. Con-

> It could probably be shown by facts and figures that there is no distinctly native American criminal class except Congress.
>
> —Mark Twain

sequently, it seems quite naive to seek an organic basis for something that is in part defined culturally. Killing a person is a criminal act, but if the person is a soldier in the army of an opposing nation in a time of war, it may be an act of heroism and patriotism.[13]

THE GENETICS OF DEVIANCE

It is not really the act itself, then, that is the subject of the scientific investigation of criminality, but rather the act in a specific context. It's not Why do people kill? but Why do people kill when they are not supposed to?

This is rather a different question. It is a question about following the rules. Why do some people follow them and others not? Is it just their nature?

The simple answer to why people do not follow rules is that they do not think the rules are fair: that in playing by the rules they cannot attain their goals. Breaking the rules then also involves a consideration of the possible penalty relative to the possible reward. And their upbringing and life experiences dictate what kinds of rule-breaking they can live with. To someone inured to violence, rape or murder may be conceivable, while blasphemy and draft-dodging may be inconceivable. To someone else, insider trading may be conceivable, but armed robbery inconceivable. The prospect of a jail sentence may be daunting for some, but not for others.

Thus, to talk about a constitutional basis for crime as if it were mono-lithic and objective is quite misleading. Rather, it is a variant on an old theme in human relations. It is an old question dressed up in new clothes, the question Why are they different from us? It is the question asked of the newcomers by the people already safely entrenched in the middle of the social hierarchy. It is a question generally of deviance: Why are *they* like that? Most specifically it is a question about morals, why some people don't hold the set of values that they are supposed to—according to the person formulating the question.

The primary group being defined here is not racial, except to the extent that race may correlate with criminality, as it has tended to be cor-related in American history. This is presumably why Frederick Goodwin (Chapter 12) was called a racist for advocating the exploration of a bio-logical basis for crime.

Yet the same arguments continuously arise for other behaviors, with a common theme. Why do women behave the way they do? Is it in their makeup, or is there some other explanation for it? Where does homo-sexuality come from? Is it constitutional, or is there another explanation? The questions are always framed in reference to perceived deviant or abnormal behavior. Normal behavior, defined narrowly, is self-evident: The problem is, What's wrong with the people who don't behave that way, like a law-abiding, heterosexual man?

Though we are not breaking up the species by specifically racial criteria any more, we are neverthe-less asking the same kinds of ques-tions that the earliest racial theo-

> You cannot teach a crab to walk straight.
>
> —Aristophanes

rists did. Given that we can divide our species up into discrete groups with behavioral differences, are those behavioral differences, which are manifestations of moral and ethical differences from "normal" Euro-American heterosexual men, constitutional?

Obviously a simple answer to any of these questions implies social policies. Is the problem organic, or not? With respect to racial differences in behavior, acculturation studies have long suggested that whatever differences in behavior are visible between "racial" groups have a predominantly cultural-historical basis. With respect to criminals, we know that generations of immigrants and the urban poor have had the privileged classes close doors in their faces, bar their upward mobility, and then blame them (their gene pools or their brain hormones) for reacting to the situation. We don't know whether criminality has a genetic component. One thing is clear, though—while the explanation doesn't change, the communities to which it is applied do. Only a few decades ago, it was the Jewish, Italian, and Irish communities whose anti-social criminalistic tendencies raised the question of whether it was inbred or acquired. Then they became upwardly mobile, and on entering the middle class, the issue of their constitutional defects became moot.

THE HEREDITARIAN JUMBLE

Like the question of just how many races of the human species exist, the question of the nature-nurture origins of human behavioral diversity is largely unanswerable until it is reformulated. This reformulation involves breaking it down into smaller and different questions, which may be answerable. The first new question is, Are observable behavioral differences between natural human populations genetically based? The answer seems to be No. The second new question is: Is moral deviancy constitutional?. This must be addressed empirically, in terms of what specific kinds of deviancy are the subjects of analysis, but seeing the common thread running through these ostensibly scientific questions highlights the historical similarities between the ways in which these questions have been approached.

The study of intelligence and its heredity was initially an attempt to give a biological explanation for the cultural dominance of Europeans. Intelligence and morality were tightly bound to one another, as (1) intelligent people are moral and (2) the primitive, unintelligent, peoples outside of western Europe are immoral. Obviously the Europeans were smarter than the peoples they had subjugated, but was this assumed intellectual superiority manifest as organic difference, and thereby more scientifically accessible? As we noted in Chapter 7, such an overly mechanistic approach to human behavior led to extensive studies of the subtle ways in which skulls vary across populations; but skull variation really wasn't the cause of cultural domination, just a biological rationalization of it.

In England, where the population was more homogeneous than in America, the emphasis was on class more than on race. Consequently, racial differences in intelligence were less significant than the demonstration of the intrinsic intellectual-moral superiority of the upper classes. The demonstration of the organic, hereditary nature of intelligence was here carried out using studies of identical twins as a natural experiment. Given their genetic identity, if identical twins were raised apart, all the variation in their intelligence should have been attributable to the environment. If identical twins reared apart were more similar to one another in their intelligence tests than pairs of other people raised apart, this would be evidence for a genetic component to intelligence. And indeed, this is what Cyril Burt found, on his rise to pre-eminence among British psychologists: The privileged classes deserved to be privileged, because they were smarter than the unprivileged, because intelligence was for the most part hereditary.

Half a century later, however, a number of anomalies in the data were discovered by Princeton psychologist Leon Kamin, and more bizarre personal and professional quirks of Burt's were uncovered by a journalist named Oliver Gillie. The major scientific anomaly is that the coefficient of correlation between the IQs of the identical twins did not change at all over a 30-year span and a tripling of the sample size. All Burt's data were destroyed upon his death, an uncommon practice among scientists, and certainly unexpected for a scientist who believes his data to be convincing. Further, Burt was known to write reviews pseudonymously (so that it would look as if other people agreed with him), and the two assistants credited with helping him collect the twin data apparently never existed.[14] All of which sent the study of twins back to the drawing board.

Other studies correlating the IQs of biological parents, adoptive parents, and children turn out to be statistically very messy, with little consistency evident in the data. For example, a study in Texas found a slightly higher correlation between a mother and her biological child than between a mother and her adopted child, while a study in Minnesota found the opposite.[15]

Certainly the most outrageous of the new twin studies is a widely publicized study reuniting twins that had been separated since birth, and separated sometimes for decades. Here, however, the "scientific" results are usually reports in the mass media, with confusing intentions and effects. For example, *Newsweek* in 1987 related the cases of Jim Springer and Jim Lewis, identical twins reunited 48 years after being separated at four weeks of age. In addition to having the same name, they had married and divorced women named Linda, remarried women named Betty, given their sons the same name, and given their dogs the same name, in addition to many other uncanny similarities.

The problem here is that this is no longer an experiment concerning heredity; it is now a mass-media fantasy about the psychic powers of twins. No geneticists in their right mind think that what name you give your dog is under the control of your genes. So what this story is about is not the concordance in intellectual performance of genetically identical people, but rather, their ESP. If their concordance of IQ and personality testing, on which a genetic argument is based, is as real as their psychic powers, then the argument is not going to be very compelling to a scientific audience. Further, quite obviously if they are in psychic contact, the psychological examinations they have been given are valueless—since they were able to cheat, and share the answers with each other!

Obviously they aren't doing that, though. The twins don't have psychic powers, nor does anyone else. But the striking similarities put forward in the media invite the public to draw precisely those conclusions.[16] The lack of credibility in this anecdotal research means that twin studies are still of highly dubious value in determining simple answers to questions concerning the inheritance of mental processes—and especially concerning the differences that may exist between groups. Recent adoption studies of non-twins find effects of heredity and effects of upbringing on within-group variation in IQ. But their relevance to the behavioral differences among groups—which is presumably what we are interested in—is minimal. Of somewhat greater interest is the study of between-group variation, showing a seven-point increase in IQs of Japanese relative to Americans over the course of a generation, apparently unrelated to genetics.[17]

So the failure to distinguish between the properties and causes of within- and between-group variation is as much a problem now as it was in the time of the eugenicists, largely because of the confusion engendered by framing the question poorly. Is intelligence inherited? is fodder for ideologues; only when it is broken down do answers emerge. Intelligence, in its various forms and accessibility to measurement, is a phenotype, which like all phenotypes comes from both genes and environment; both contribute significantly to the variation in intelligence within groups. Between groups, however, environment accounts for the vast majority of the variation.

Why should this be the case? Even if we (completely gratuitously) assume intelligence to be monolithic and performance to be a reliable measure of ability, the cryptic genetic variation for intelligence should be patterned in much the same way as the rest of the genetic variation known in the human species. This would imply that most populations have most alleles, and the only significant variations are quantitative. If we then consider the heterogeneous nature of intelligence, the difficulty in establishing small differences in it between individuals, and the large impact that environmental variation has upon it, there seems very little reason to think that whatever variation might exist in intelligence between groups would most reasonably be explained by recourse to genetics.

The hereditarian case, then, rests on the assumptions that (1) variation in intelligence can reliably be ascertained, (2) genetic variation makes a significant contribution to it, and (3) the pattern of hypothetical genetic variation in intelligence in the human species is different from the known patterns of genetic variation. Further complicating matters is the old problem of establishing the range of normal variation from the study of pathology—especially in psychological traits.

The genetic basis of aggression was discovered in 1965, for a time. Finding a higher proportion of men with XYY chromosomal constitutions in mental/penal institutions than predicted by chance, the geneticists hypothesized that (in concert with cultural stereotypes) the men were there because they had higher levels of stupidity and aggression from having essentially a double-dose of maleness—two Y chromosomes instead of the normal one.[18] The immediate difficulties, however, involved the fact that (1) XXYs were also over-represented, (2) there were far more XYYs at large than in prison populations, and (3) this would account, in any event, for a minute proportion of violent crime and aggression.

Follow-up work has led to a different interpretation than the original one. The human body is strongly constrained to have two, and only two, copies of each chromosome or chromosome part. Having an additional chromosome almost invariably results in the death of the embryo or fetus; the exceptions involve the smallest chromosomes, or small regions of chromosomes, with very little genetic material.[19] Having an extra copy of any chromosomal region then results in various pathological phenotypes whose common thread is mental retardation.

The XYY phenotype involves tallness, bad skin, and slight retardation. In other words, these people are often threatening-looking, stand out, and are easier to apprehend than average. There is no evidence that they are more aggressive than ordinary people, or that specifically the extra Y-chromosome (as opposed to just chromosomal material generally) leads to their over-representation in mental/penal settings.[20] Nevertheless, public fascination with a simple organic cause of aggression was fueled by erroneous reports that mass murderer Richard Speck had an XYY constitution. More recently, the action of the science fiction movie sequel *Alien3* takes place on a planet inhabited by the most dreaded criminals in the galaxy—the XYY men.

Mapping psychological variation directly onto genetic variation is a simple and sweet explanation, in harmony with cultural stereotypes—

but not in this case the one held by the genetics community. Normal genetic development is quite complex: the same genotypes often result in different phenotypes, and different genotypes often produce the same phenotypes.[21] The direct translation of genotypic variation into phenotypic variation, therefore, is an explanation deriving from, and most suited to, the study of pathological variation—genetic disease.

We saw in Chapter 9 that the relationship between pathology and normal variation is obscure, for learning the myriad ways in which a physiological system can break down may not tell you much about the ways in which it operates or varies in nature. Advocates of strong hereditary links to human behavior often confuse within-group variation, between-group variation, and normal-versus-pathological variation. These may have nothing to do with one another.

This is a mistake made by the eugenicists in targeting "feeblemindedness," which they defined very broadly and in opposition to implied normality, which they imagined to be quite narrow. Thus, the study of variation *became* the study of pathology. Further, this concealed the fundamental issue, which was that of morality: These foreigners did things that were loathsome, and it was presumably on account of their bad brains. The eugenicists wanted to know: (1) How much behavioral variation is tolerable? (2) Where does it come from? and (3) What can we do about it? The answers they came up with were (1) very little; (2) the genes; and (3) eliminate or sterilize the bearers.

Again, there are several different and better questions to be asked here than the false question, Is behavior genetic? These are: What is the acceptable normal range of behavior? On what basis is it established? What is the relationship between *deviation from* that range and *variation within* that range? Does variation within it have a genetic component? If so, is there within-group genetic variation for it? Is there between-group genetic variation for it? Until these questions are unjumbled, the strong hereditarian stance is one that has little scientific merit, though it affords simple answers to social problems, as history shows.

THE GENETIC BASIS OF SEXUAL DEVIANCE

One of the parallel avenues of speculation is on the nature of the differences between men and women. Like the difference between ax murderers and Harvard professors, the difference in behavior between a boy and a girl may tell us nothing at all about the differences between boys and girls. Differences between men and women are genetic (biological), developmental (at the individual level), and cultural (at the group level).

Can we allocate specific behavioral differences between men and women to these categories, and specifically to the first category?

There is no way to conceive of an experiment that would control adequately for all the variables present. The arguments in favor of a hereditary determination come from two main sources: performances (either on tests, or observational data) and brain structure. As we have already seen, variation in performance does not necessarily imply variation in ability—the elusive metaphysically biological construct at the bottom of this question.

The second is well-established anatomically, that the brains of men and women differ, on the average.[22] Of course, the rest of women's bodies are also different from the rest of men's bodies—though few organ systems have been studied so comprehensively as the brains have in attempts to find such differences. Given that average brains differ between men and women, is that what causes women to act like women?

This is a thorny question, whose deterministic assumptions about anatomy and behavior lie at the very root of the hereditarian tradition. Again, however, we have encountered it before. Because the brain is the seat of the mind, and the mind is composed of thoughts, it follows that different brains cause different thoughts. But do they? We know something about the workings of "the brain," and something about the effects of gross pathologies upon it. But what does that tell us, if anything, about the effects of more subtle variations relative to the brain's normal functions?

There is considerable cultural baggage again built into this line of argument. Is the structure of the brain an independent variable in behavior, causing thoughts and deeds, but not reciprocally caused by them? The brain develops interactively throughout the course of an individual's growth, and its fine structure is not genetically pre-determined; so it is not a genetic "given" in the equation.[23] And, of course, this is the same hyper-materialism that placed the study of skull size and shape at the forefront of racial studies for nearly a century.

Like most differences between populations of people, there is a considerable overlap in the brains of men and women. What, then, does this mean with respect to determining their differences in behavior? Does it mean that people should have a CAT scan to determine their occupational status and salary level? Or should it be based on their level of skill and job performance, independently of their brain structure?[24] Like the racial studies, the study of women's brains in relation to men's brains drags a social program along with it. If their behavioral differences are intrinsic, then (like the unfavored races in relation to the favored, and like criminals in relation to the law-abiding) there are good reasons for them to be treated as they are, and attempts to equalize or normalize their treatment, or improve their lot, would be at best unnatural. After all, you can't make a silk purse out of a sow's ear.

This line of research is clearly not value-neutral. It would take a naive scientist, indeed, to imagine that this research exists in a social vacuum. A valuable lesson from eugenics applies here: One needs to think very carefully about the chain of reasoning and interpretation of conclusions from this line of research. It affects people's welfare. Research about the intrinsic nature of behavioral variation between groups needs to be controlled and documented far more rigorously than research involving flies or clams. Unfortunately, history also tells us that scientists are willing to draw grand conclusions far more readily in this area than they would if they were studying flies or clams.

Likewise, the "genetic basis" of homosexuality. Here the evidence rests on two widely publicized kinds of studies: brains and twins—the epistemologically weakest that the history of human biology has to offer. The brain study, by neuroanatomist Simon LeVay, found a difference in the size of the third hypothalamic nucleus between a sample of 19 male homosexuals and heterosexual male controls, the nucleus of the male homosexuals being smaller than the male heterosexuals, and like the females.[25]

Once again, we find heavy cultural loading of ostensibly objective scientific work. We begin with the assumption of the brain structure as being an independent variable in human behavior. We progress to the fact that female homosexuals were not studied, though presumably they would have provided a critical positive control: If lesbian brains are like men's brains, that would accord with the hypothesis; if not, that would suggest that either the hypothesis of an intrinsic nature of homosexuality is flawed, or that male and female homosexuality may have different causes, and are not simple reciprocals of one another. And we finish with a basic reification: confusing the properties of words with the properties of the things they stand for. The fact that we have two contrasting words, homosexual and heterosexual, does not mean that there are two kinds of people in the human species. The spectrum of human sexual behavior is very broad, both at the individual level and at the cultural level. The division of humans into two kinds, homosexual and heterosexual, reflects a specifically cultural value about the artificially dichotomous nature of human sexuality. Further, the idea that even male homosexuality is monolithic is quite unrealistic, given the broad range of feelings and behaviors among gay men.[26]

Returning to the issue, then, is this deviant behavior intrinsic or extrinsic? Historically, there have been strong social pressures for a clear-cut answer. In the early part of this century, the ostracism of homosexuals, most extremely by the Nazis, provided a strong impetus to argue that homosexuality is not a constitutional feature, but a learned behavior. Since one could not then eliminate it, as a learned behavior would always be present, the reasonable solution would be to come to a mutually tolerable co-existence with homosexuality. This attitude essentially prevailed until the "homophobes" discovered homosexuals teaching in schools. Fearful that their children were being taught buggery instead of algebra, they sought to bar homosexuals from teaching—thus creating a strong impetus to argue that homosexuality is not learned, but innate.

GENETIC BEHAVIOR: HERE TODAY, GONE TOMORROW

Another group of scientists claimed in 1993 to have located, though not identified, a genetic basis for homosexuality, at the tip of the long arm of the X chromosome. Finding a suggestion of maternal inheritance in some of 114 families of homosexual men, they carefully chose 40 to study at the molecular level. They found an association between five genetic markers and 33 of 40 sets of homosexual siblings. This association is statistical, not mechanistic—the genetic markers are not genes for homosexuality, simply variable bits of DNA, around which a "gene for homosexuality" may lie.

Since the subjects for the DNA marker study were carefully selected, we have at best here a potential explanation for some subset of specifically male homosexuality. There is no suggestion of a mechanism, no attempt to explain *most* male homosexuality or *any* female homosexuality: just a concordance of genetic markers in 64 percent of a specially selected sample of homosexual brothers. Yet *Time* magazine queried "Born Gay?" and expounded: "Studies of family trees and DNA make the case that male homosexuality is in the genes."[27]

Once again, the scientific issues are subservient to the social issues, and whatever objectivity the science ever possessed is lost in a sea of social advocacy and validation. Like the eugenics debate, the center of the argument is the nature of morals and values: what causes *them* to be different, and ultimately what can or should be done about it. Certainly the words of geneticist Thomas Hunt Morgan, in one of the earliest criticisms of eugenicists from an American biologist, bear noting:

> [I]t is not so much the physically defective that appeal to their sympathies as the "morally deficient" and this is supposed to apply to mental traits rather than to physical characters.[28]

What genetics tells us is that there is a broad range of genetic variation in all populations; that the alleles enter into a broad range of combinations; and that these combinations are manifest as a broad range of phenotypes. That some of the variation might be affecting behavior is certainly possible, but the genetic variation affecting behavior is probably identical from population to population, and thus has no part in explaining group differences. Further, in studying only deviant behaviors, and analyzing them as if they were

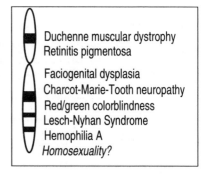

Duchenne muscular dystrophy
Retinitis pigmentosa
Faciogenital dysplasia
Charcot-Marie-Tooth neuropathy
Red/green colorblindness
Lesch-Nyhan Syndrome
Hemophilia A
Homosexuality?

Figure 13.2. Some genetic syndromes mapped to the X chromosome.

genetic pathologies, the science being done is far from being value-free. It may be worthwhile to contemplate the implications of "homosexuality" lying alongside the genes on the X chromosome already known to cause diagnosable anatomical phenotypes— all of which are diseases (Figure 13.2).[29]

All this discussion, of course, assumes that the linkage between the X-chromosome region and the 64 percent of the selected homosexual brothers holds up. Actually several other complex "abnormal" behaviors have been recently reported to be linked to chromosome segments, only to be subsequently refuted or retracted: alcoholism, schizophrenia, and manic-depression.[30]

The "gene for homosexuality" must also be seen in the light of two recent studies, each purporting to have identified a genetic basis for a specific behavior. Again the behavior is definitionally abnormal—in this case, hyperactivity on the one hand and aggression on the other. But the hyperactivity gene accounts only for a tiny fraction of the hyperactive children at large, as the authors acknowledge. Why? Because the phenotype is so heterogeneous that it subsumes several different natures and causes. The "gene for hyperactivity" in fact accounts for very little hyperactivity. Likewise, the "gene for aggression" was found in a Dutch kindred, several of whom were spousal abusers and has a number of clinically pathological phenotypes (including retardation) associated with it. That it would account for any significant proportion of aggressive acts in the population at large—assuming we could even delineate them adequately—is ridiculous. Nevertheless, it was touted in the press as a "gene for aggression."[31]

The basic problem lies in extrapolating from rare behavioral pathology to common deviant behaviors. Lesch-Nyhan syndrome (Chapter 8) probably tells us nothing about nail-biters; the allele segregating in the Dutch kindred probably tells us nothing about the "wilding" of the Central Park Jogger. It is largely an artifact of our lack of knowledge of how genetic systems work, and of relying heavily instead on their rare pathological breakdowns to infer normal function.

Given that extensive genetic diversity and phenotypic diversity do not map very easily onto one another, why is there so much attention focused on finding genes for specific deviant behaviors? It isn't unthinkable that something as heterogeneous as sexuality would have genetic variation associated with it. The possibility, however, that a simple genetic system explains such profoundly complex phenotypes is remote at best. The extraordinary aspect of these arguments is that although the technology has changed, the mode of explanation hasn't progressed in over a century: It's there, it's abnormal, it's innate.

PLATONISM AND THE SEARCH FOR HUMAN NATURE

Certainly the most tenacious assertions about human behaviors come from the study of human nature. Here, the arguments are not about constitutional differences underlying the behavioral differences among human populations, but rather about the basic constitution of the human mind, the genetic roots underlying human nature.

Humans are basically territorial and aggressive, argued the playwright Robert Ardrey in a series of books in the 1960s. Humans are basically hypersexual "naked apes," argued ethologist Desmond Morris.[32] Though this may sound a bit like science, the arena has shifted subtly toward the legalistic, where argument is the currency, and evidence is hoarded and traded to fit the needs of the argument. Experimental controls are unimportant, and experiments themselves are rarely of any interest, except as they can be invoked, usually irrelevantly.

This is not to say that such literature is inferior, for it is often exciting and provocative (frankly, unlike most scientific work), only that it is not science, although it is often made to look like science, and sometimes comes from scientists. After all, if an argument is scientific, it has greater social power than it would otherwise command, as the eugenicists knew quite well.

> He who says there is no such thing as an honest man, you may be sure is himself a knave.
>
> —George Berkeley

Humans, for example, are by nature somewhat polygamous, declare some popular works on the "scientific" basis of human nature.[33] After all, among Old World higher primates, there is a loose correlation between mating system and sexual dimorphism in body size. In the chimpanzees, gorillas, and baboons, where a male has sexual access to several females, males are also considerably larger than females. Among the gibbons, where a single male and female are pair-bonded, and live in the forest canopy with "traditional family values," males and females are the same size. Human males average about 20 percent larger than females. Does it not then follow that we are constitutionally somewhat polygamous?

As we noted in Chapter 2, this argument focuses on a single variable, ignoring other differences in sexual dimorphism between humans and our close relatives. Notably, those polygynous species, in addition to differences in overall body size between males and females, have large differences in the size of their canine teeth. Human males and females do not. This implies precisely the opposite: that humans are like monogamous primates. More to the point, though, humans have patterns of sexual dimorphism unlike our close relatives: in body composition, body hair, and facial hair, most significantly. These have no parallel in our close relatives, and suggest a more complicated state of affairs—that we cannot extrapolate directly from the sexual dimorphism of our relatives to our own, because of something unique in the ancestry of humans.

Pronouncements on human nature sometimes use cross-cultural data to support their position. For example, a larger number of societies have been polygamous than strictly monogamous. But is it fair to characterize the constitution of individuals by the ethics and values of their culture? Again, the eugenicists thought so in the 1920s. But obviously there are societies in which polygamy is acceptable and those in which it is not, and history tells us that without significant change in the gene pool, there can be extensive change in the mating patterns. And human behavior is sufficiently complex that there are societies in which polygamy is acceptable, but only practiced by very few (those who can afford it). Is this society polygamous (because it is acceptable) or monogamous (because it is the most common practice, with the polygamists being deviants)?

Is not even clear what should constitute polygamy—how we would recognize it when we see it. Literally it means having more than one spouse at the same time; but what about having one spouse and several legal concubines, as in ancient Rome? Or having one spouse and other love affairs? Or having one spouse, no love affairs, but sexual fantasies about other people? Or having several legal spouses, but relations with only one? Or having only one spouse when you are permitted more? It is not at all clear that human behavior permits us to make such a simple diagnosis as the simplicity of the words we use might suggest.

Unsurprisingly, pronouncing humans to be constitutionally polygamous stands to validate some cultural stereotypes, particularly the one in which men have wild oats to sow indiscriminately, and women remain in the nest, saving themselves for "Mister Right." Perhaps this is our nature. There is certainly no way to subject it to direct experimentation. But it raises an interesting question about the nature of deviancy. What about women with wild oats to sow, and monogamous men? Are they now definitionally not human? Are they mutants? How do we account for them?

> Oh, gallant was the first love, and glittering and fine;
> The second love was water, in a clear white cup;
> The third love was his, and the fourth was mine;
> And after that, I always get them all mixed up.
>
> —Dorothy Parker

More to the point, this assertion about human nature is fundamentally ahistorical. After all, the social world is constantly changing, and the laws and values that exist in one time and place have not always been there. How do we account for the fact that, regardless of what cultures have tolerated in the past, polygamy is illegal over most of the world today? Laws in opposition to human nature are notoriously short-lived and narrowly applied.[34]

Where, then, are we with respect to polygamy as human nature? It has a dubious basis in primate biology. It ignores human history. It confuses individual behaviors with cultural rules. And also, by the way, it serves scientifically to validate cultural stereotypes.

What we have here is not science, but a philosophical trap laid by Plato, and sprung most notoriously by the eugenicists. It is the idea that there is a unitary human nature that can be encapsulated in single words or phrases. This implies a very narrow definition of what is human, or more realistically, what is *normal* for a human. The major scientific revolution in the study of the human species in the latter half of the 20th century, however, has been to undermine that idea. It has been the demonstration that human nature is highly heterogeneous: that humans are very diverse behaviorally, from group to group, with no objective way to distinguish which groups are behaving "naturally" or "unnaturally."

To argue that one's own values represent human nature is quaint and dull. People have done it ever since they have argued about what constitutes human nature. Behaviors and values among human groups are exceedingly diverse, for (as far as we can tell) historical and not genetic reasons. Further, within each group, the genetic constitution of individuals is highly heterogeneous. The assumption of a universal, fixed, and narrow human nature implies genetic homogeneity for whatever loci underpin that nature. We saw in Chapter 9 that this underlies the original conception of the Human Genome Project, but is empirically untenable. In the genetic systems available for study we find empirically large amounts of genetic diversity. Obviously, if we can extrapolate from this (which is all we have) to the genes controlling human nature, it would imply heterogeneity of human nature as well.

Thus, we have consistently found extensive *behavioral heterogeneity between groups* and *genetic heterogeneity within groups*. Humans are thus behaviorally highly polytypic and genetically highly polymorphic. This is not what would be expected under the assumptions of a narrow range of human nature, nor of significant genetic control of its variation across the map.

Humans seem to be capable of legitimizing a broad range of mating patterns: in one time and place polygamy and monogamy, in another only monogamy, and in still another polyandry. Within each culture, there are people who practice one and not another; there are people who do things they are not supposed to do; and people who only dream about doing things they are not supposed to do. What this implies is a heterogeneous human nature, not a fixed allele in the gene pool.

WAS HAMMERSTEIN WRONG?

Librettist/lyricist Oscar Hammerstein II included in *South Pacific* a song that spoke for a generation coming to grips with World War II. Is intolerance basic to human nature? No, sang Lieutenant Cable, speaking for the author: "You've got to be taught to hate and fear . . . you've got to be carefully taught."

Hammerstein's assertion, however, has been implicitly rejected by some recent writers on human nature, who maintain that, like polygamy, "xenophobia" (or fear of strangers) is part of human nature. According to author Jared Diamond, chimpanzees are genocidal and "also share xenophobia with us." "In short," he adds, "of all our human hallmarks—art, spoken language, drugs, and the others—the one that has been derived most straightforwardly from animal precursors is genocide."[35]

In this argument, "xenophobia" is a plesiomorphic constitutional endowment of our species,[36] expressing itself most lethally as genocide, which is documentable over many lands and times. But the same set of questions emerges. Is there genetic variation for it? There is certainly phenotypic variation for it. Are deviant nice people mutants? Is the genocide of Native Americans by Spanish conquistadors or American soldiers an expression of the nature of the people composing the army, or simply the enforcement of state policy in that time and place?

It is trivially obvious that humans are capable of genocide; after all, genocide has occurred. That means people were capable of it. But is it human nature, or merely one of the extraordinary things humans and their institutions are capable of doing? If genocide, like cheating on your wife, is "just human nature," then to what extent is one accountable for one's morally deviant actions? What we have here is not so much a justification for these deplorable acts, as a trivialization of them.

> When the last red man has vanished from this earth, and his memory is only a story among the whites, these shores will still swarm with the invisible dead of my people. . . . The white man will never be alone.
>
> —Seattle

Xenophobia is widespread in humans, but it is a peculiar sort of xenophobia. To say merely that it exists is to assume that there is a natural difference between the group and the feared-loathed stranger. As we noted in Chapter 10, differences between human groups are largely constructed, not natural: they are based on language, dress, custom—cultural features that permit self-definition. In other words, Red Sox fans may hate Yankee fans passionately, and it is fun to do so (though English soccer fans are better known for acting out their animosities), but it really is quite silly. And further, most Red Sox fans will gladly join forces with Yankee fans against Dodger fans. Anthropologists call this a segmentary lineage—a hierarchical model of alliance.

To return to xenophobia, then, humans are xenophobic about incredibly inane things. Some are xenophobic about what people look like, other are xenophobic about the name of the deity others pray to, and still others are xenophobic about other people's occasional leisure activities. What does this tell us about the Tasmanians at the hands of the Europeans, or the Jews at the hands of the Germans? Asserting that these were instances of "human nature" doesn't tell us anything interesting about the causes of these episodes, nor of the episodes themselves. It can function as a biological excuse—not necessarily a justification, but an excuse—taking the spotlight off the actor and the actions, and redirecting it to the germ-plasm.

The weakness of the "xenophobia/genocide-as-human-nature" idea is that it takes the existence of between-group variation as given. It assumes natural, objective differences between the victim and perpetrator in reconciling their antagonism to our genome. Actually, however, those differences are constructed culturally, which is a fascinatingly human characteristic. The point is *not* that we have a drive to hate people different from ourselves; it's that we *define* ourselves culturally, and then *make* people different from us. Would the Nazis have been more humane if there had been no Jews? Certainly not; the Jews were there, but the role they played for the Nazis was a construction of Nazi culture. If the Jews didn't exist, they would have been invented (as to a large extent they were!), or somebody else would have filled the bill.

Irish Catholics and Protestants hate each other, yet are virtually the

same, likewise Bosnians and Serbs, and Tutsi and Hutu. Where antagonistic human groups are biologically distinct from one another, it may
appear that the xenophobia is based on a natural difference. But the fact
that the same thing happens between groups *lacking* significant biological differences shows that the interesting aspect of this behavior is
strictly cultural.[37]

The lesson of the Holocaust, then, is not that it is another expression
of our inhumanity. That serves merely to trivialize it. The lesson derives
from the fact that it occurred at a time in which Europeans were thought
to have achieved some degree of enlightenment, and that it was carried
out by Europeans against themselves. It is the cultural construction of
differences between human groups that we take from the Holocaust;
how easy it is to mistake these for constitutional differences, and how
scientists can be the intellectual leaders of the confusion.

Whether it is an instinctive attribute of human nature to hate and fear,
or you have to be carefully taught it, is a poor question. *Whom* you hate
and fear is learned; and *why* you hate and fear them, and *what* should
be done about them, are acquired knowledge. That is what the scientific
study of human biology and its history demonstrate, and what the focus
of xenophobia should be.

RACE, XENOPHOBIA, AND THE LESSONS OF HISTORY

The fallacy of pointing to episodes of genocide in human history and
exclaiming, "Look, it's human nature!" is that (1) it is trivial to note that
genocide is within our behavioral capabilities; (2) it only happens infrequently; and (3) it confuses the goals of cultural institutions with the
motivations of their agents. Most importantly, however, to do so is to
rediscover the conceptual error of the era of racial anthropology. That
error consisted in taking the continuous nature of human variation,
examining differences between various extremes, and concluding that
the major source of diversity within the human species was a fundamentally natural set of boundaries among large human groups.

No such set of boundaries exists. The differences among human
groups are for the most part differences of self-identification, using categories that are culturally constructed. Sometimes they correlate with
biological differences; often they don't. The basic error lies in confusing
the cultural boundaries for natural ones, and then concluding that the
groups they delineate are real, rather than constructed.

Thus we come full circle in the study of human diversity. We began
with "race," a division of our species, each with its unique biological
constitution, as the focus of the science of human variation; and we have

ended with "xenophobia," a genetic quality that disposes you to hate members of groups constitutionally different from yours.

Both are wrong, for the same reason. It is human nature to create divisions where none exist; to classify things that defy classification; to impose a semblance of order on what would otherwise be a formless jumble of sensory impressions; and to extract meaning from that the order. This order is culture, and it is, so far as we can tell, one of the fundamental ways in which humans differ biologically from our close relatives, the apes.

Race was one way of ordering human diversity. It doesn't work because the biological differences between human groups are trifling compared to those within the groups, and because the major biological divisions of humans presumed to be "out there" do not manifest themselves clearly. Race doesn't explain the patterns of diversity of human behavior; and ultimately even simple classifications of races emerge to be based more on cultural perceptions of who-is-more-like-whom than on biological criteria.

Those same criticisms undermine the utility of taking xenophobia as human nature. It assumes there is someone basically different, out there, to hate. But there need not be. Those who are inclined toward group animosities create such groups themselves. Like perceptions of race, they can be augmented by the presence of biological variation, but they are not driven by it. Biological difference is neither necessary nor sufficient for groups to perceive themselves as different and for animosities to exist between them.

NOTES

1. Selvin (1991:368).
2. Mencken (1927).
3. Kottack (1985).
4. Klein (1991).
5. The light-heavyweight champ from 1916 to 1920 was called Battling Levinsky; from 1930 to 1934 it was Maxey Rosenbloom. The lightweight champ from 1917 to 1925 was Benny Leonard; the featherweight champ from 1925 to 1927 was called Kid Kaplan. Sachar (1992:352–53) discusses the eminence of Jewish boxers in the early twentieth century. The last prominent Jewish boxer was Mike Rossman, who held the WBC light-heavyweight title briefly in 1978. In basketball, at the time of this writing, the only Jewish player in the NBA is Danny Schayes of the L. A. Lakers.
6. Ashe (1993a:47). See also Olsen (1968).
7. Ibid., p. 43. See also Kareem Abdul-Jabbar, quoted in Lapchick (1989).
8. Ibid., p. 44. See Cobb (1936, 1947), and especially Wiggins (1989). I thank John Hoberman for directing me to this literature.

9. Ashe (1993b:22).

10. Ibid., p. 50.

11. Hooton (1939b:308).

12. Ibid., p. 309.

13. Merton and Montagu (1940). See also the review of Wilson and Herrn-stein (1985) by Kamin (1986); also Young (1989).

14. Dorfman (1978) and Hearnshaw (1979) review the evidence for falsifica-tion of data on Burt's part. Joynson (1989), Fletcher (1991), and Jensen (1992) attempt to defend Burt's research as merely sloppy, stupid, strange, and sad—but not fraudulent—and to criticize the critics as ideologically motivated. Jensen holds out the hope that Burt's two researcher assistants may have actually existed, but acknowledges that " as his most notable eccentricity, he wrote a con-siderable number of articles, mostly book reviews (it remains uncertain just how many), under various pseudonyms or initials of unidentifiable names" (p. 101). This does not resemble the behavior of ordinary, responsible, honest scientists.

15. Scarr and Weinberg (1978); Horn, Loehlin, and Willerman (1979); Lewon-tin, Rose, and Kamin (1984).

16. Begley, Murr, Springen, Gordon, and Harrison (1987). A subheading of this article was indeed given to "ESP events." See also Lang (1987), Bouchard et al. (1990), Dudley (1991), Beckwith et al. (1991).

17. For the adoption studies, see Teasdale and Owen (1984), Capron and Duyme (1989). For "the great Japanese IQ increase," see Lynn (1982), Anderson (1982).

18. Jacobs, Brunton, Melville, Brittain, and McClemont (1965).

19. Or, in the case of the X chromosome, ordinarily deactivated, such that only one copy typically is functional in a cell (Lyon, 1992).

20. Hook (1973), Witkin et al. (1976).

21. Lerner (1954) emphasized the difficulty in reasoning from genotypic to phenotypic variation. Dobzhansky (1970) discusses a "norm of reaction" set by a single genotype, resulting in many possible phenotypes. Waddington (1942, 1957, 1960) pioneered the appreciation of the "buffering" or "canalization" of the phenotype, wherein normal phenotypes are stably produced in spite of genetic variation.

22. Lacoste-Utamsing and Holloway (1982), Swaab and Fliers (1985), Kimura (1992).

23. Kandel and Hawkins (1992), Shatz (1992).

24. Hines (1993).

25. LeVay (1991). For the twin study, see Bailey and Pillard (1991), Holden (1992), Manoach (1992).

26. Carrier and Gellert (1991), Ehrenreich (1993).

27. Hamer, Hu, Magnuson, Hu, and Pattatucci (1993), Pool (1993), Henry (1993), Fausto-Sterling and Balaban (1993), Diamond (1993), Risch et al. (1993).

28. Morgan (1925:205).

29. For the general issues of linkage studies, see Risch (1992), Horgan (1993). For manic-depression, Egeland et al. (1987), Hodgkinson et al. (1987), Robertson (1987, 1989), Kelsoe et al. (1989), Barinaga (1989). For alcoholism, Devor and

Cloninger (1989), Horgan (1992), Holden (1991). For schizophrenia, Byerley (1989), Detera-Wadleigh et al. (1989).

30. In a recent review of human behavioral genetics, Plomin et al. (1994) list as "Reported linkages and associations with complex behaviors" the following: mental retardation, Alzheimer's disease, Violence, Hyperactivity, Paranoid schizophrenia, Alcoholism/drug abuse, Reading disability, and *Sexual orientation*.

31. For hyperactivity, Hauser et al. (1993), Holden (1993). For aggression, Brunner et al. (1993), Morell (1993).

32. Ardrey (1961, 1966), Morris (1967).

33. Wilson (1978:125–28), Diamond (1992:71–72).

34. This argument, curiously, is also the reverse of the hereditarian argument for the incest taboo. The widespread illegality of polygamy is taken as having no bearing on human nature; whereas the widespread illegality of having sexual relations with close relatives is taken as evidence of a constitutional basis for it.

35. Diamond (1992:294).

36. Though aggressive encounters between primates and between primate groups are well-documented, lethal inter-group aggression is rare among the apes. It has only been seen among the apes in *Pan troglogytes*, the common chimpanzee, and then only in a few cases. The best-known case is the extermination of the Kahama community at Gombe by the Kasakela (Goodall 1986). We do not know why they did it or what its relationship is to inter-group aggression in humans.

In chimps, inter-group violence is invariably aggressive and agonistic, but that is not necessarily the case in human warfare or genocide. There is nothing objectively aggressive about pushing the button that releases the A-bomb from the Enola Gay, or flipping the switch that releases the Zyklon B into the shower stalls at Auschwitz. Many of us perform the same actions—pushing buttons and flipping switches—all the time, but by performing the action on a *particular* switch or lever, you kill conspecifics you do not know, without exerting any physical effort or even confronting them, because your state has a policy in effect, you may agree with it, and you are obliged by virtue of your position to carry it out. And the human directing the lethal actions may not participate in carrying them out.

In the Gombe case, the chimps of the two groups "knew" each other, and the members of the Kasakela group appropriated the range of the Kahama. Apparently, resources were directly at issue, and the members of the group directly benefited by the action. The motivations of humans in war and of their leaders are very diverse (Ferguson 1990; Robarchek 1990). While there may be continuity of a sort between this chimpanzee behavior and human behavior, it is certainly overstated by the sloppy use of terms such as war, genocide, xenophobia, aggression, and violence.

37. Franz Boas noted this very explicitly in *Time* magazine's cover story on him (May 11, 1936), in opposition to Sir Arthur Keith's assertion that racism is natural.

14

Conclusions

Last week a crazed gunman terrorized hostages in a bar in Berkeley, killing one and wounding many others. Homicidal maniacs have appeared in all cultures over the entire length of human history. Society's modern response to their chaotic behavior has most often been a diligent search of their childhoods, as though understanding their upbringing and circumstances would explain their aberrant actions. There is nothing wrong with that kind of investigation, and in some cases history and environment will reveal clues. However, it is time the world recognized that the brain is an organ like other organs—like the kidney, the lung, the heart—and that it can go wrong not only as the result of abuse, but also because of hereditary defects utterly unrelated to environmental influences. Some inherent defects may be exacerbated by environmental conditions, but the irrational output of a faulty brain is like the faulty wiring of a computer, in which failure is caused not by the information fed into the computer, but by incorrect processing of that information after it enters the black box.[1]

The preceding quotation would be quaint, like those from Davenport or Hooton in the early part of this century, had it not come from the pen of Daniel Koshland, the editor of *Science*, in 1990. There is a good reason why we study history. It is the strongest weapon we have in the arsenal of self-comprehension and social improvement. Genes make brains, brains make thoughts, and faultily wired brains make people think and behave in deviant ways, like a homicidal maniac; therefore genes make homicidal maniacs. Maybe, but that is the same hyper-materialistic logic by which genes make homosexuals, Jews, and people who like rap music.

The title of Koshland's editorial was "A Rational Approach to the Irrational," the implication being that a rational or scientific approach to crime necessitates localizing it to the genes—the genes that make the deviant brains, produce the deviant thoughts, and make people act in those bizarre and incomprehensible ways. If the opposite approach to

Koshland's Is that of the irrational humanities, then it is appropriate to turn to a classic (if irrational) discipline, intellectual history, to show that the argument is old and hasn't borne fruit yet. Hereditarianism is a cultural value, independent of scientific knowledge or advances. There is, however, a correlation: where genetics advances, hereditarianism accompanies it. It predates genetics, but does draw legitimacy from genetics.

A year earlier, Koshland had written as a justification for the Human Genome Project, that its

> benefits to science . . . are clear. Illnesses such as manic depression, Alzheimer's, schizophrenia, and heart disease are probably all multigenic and even more difficult to unravel than cystic fibrosis. Yet these diseases are at the root of many current societal problems. The costs of mental illness, the difficult civil liberties problems they cause, the pain to the individual, all cry out for an early solution that involves prevention, not caretaking. To continue the current warehousing or neglect of these people, many of whom are in the ranks of the homeless, is the equivalent of providing iron lungs to polio victims at the expense of working on a vaccine.[2]

If there were only a vaccine against homelessness . . . Wouldn't we all be better off?

Diseases whose genetic factors have been intensively sought and are still very ambiguously genetic become simply multigenic. Economic problems become constitutional. The names of the conditions change, but the argument itself is remarkably resilient. Without the benefit of knowing how these problems have been addressed by previous generations of scientific hereditarians, it is easy to stand up, point at someone, and maintain that science shows their problem is in their genes.

We can't do the scientific controls to make this a scientific study. But we can observe that, with the same arguments, it doesn't seem to have been in their genes *then*, and that makes it unlikely it will be found in their genes *now*. The genes are just a simple answer, a scientific-sounding one, and one that allows the speaker to be abstracted from both the problem and its solution. While "extirpation" may no longer be an issue, the waste of scientific resources is.

Herbert Spencer Jennings was among the first American biologists to challenge some of the strongest hereditarian claims associated with eugenics. He did so as a eugenicist, but as one who was beginning to recognize the extravagance of some of its ostensibly scientific claims:

> Students of heredity, like other [people], are disposed to make the most of their achievements: to dwell upon what they know, what they can do, and what they can predict. They have, indeed, achieved much; the last twenty-five years have made greater advance in the knowledge of heredity than

had all the ages before. But recognition of limitations is as valuable as other sorts of knowledge; realization of what we cannot do is as necessary for correct guidance as realization of what we can do.[3]

His words have a timeless quality, particularly when read in the light of the social and political movements of the later 20th century. We seek a path to self-awareness through genetics, yet we are constantly led into intellectual cul-de-sacs.

The contemporary state of the science of the human species—and particularly of genetic variation within it—overturns seven long-held and widely held assumptions about the enterprise. These are the stumbling-blocks for previous generations and, unfortunately, for many popular writers on the subject even today. Each has particular implications for how we see the science, and how we carry out the science; and each has led us astray before.

1. THE IDEOLOGICAL NEUTRALITY OF SCIENCE

Where science gives validity to ideas, science is corruptible.[4] Not simply in the sense of scientists being bought by corporations to mislead the public, as the tobacco companies recognized earlier this century, but corruptible in the sense of scientists being members of a society and having the same cultural values as others. Except that the cultural values possessed or expressed by the scientist can be construed as scientific ideas, because a scientist has them.

The problem is that a scientist who is not actually a professional student of human biology may have no greater insight into the problems of human biology than the checkout clerk at the grocery store. This was one of the major problems with "sociobiology" in the 1970s—it was a series of generalizations about human behavior expounded by students of insects and birds. What they often brought to the study of human biology was the prestige of scientific authority in the exposition of very old-fashioned cultural values. In some cases this was little more than anti-intellectual dilettantism, with scientists from another discipline ignoring the progress that had been made in the study of humans, so as to approach with a fresh and naive perspective, and discover the same old mistakes all over again.

Most of the fizz left the sociobiology debate as crude hereditarianism was exorcised from it, and as the more recent sociobiology has been transformed into behavioral ecology. Its major impact now lies in formalizing decision-making strategies on the parts of cultures and individuals.

The important residual, however, is that science is a validation mechanism in modern society, and scientists bear a responsibility for what they say about the scientific analysis of our species. This lesson had been learned within the community of anthropologists decades earlier.

A major textbook on *Animal Behavior* feels obligated to defend sociobiology, by asserting that it makes no value judgments:

> Darwin's theory of evolution has been misunderstood and misused by some persons to defend the principle that the rich are evolutionarily superior beings, as well as to promote unabashedly racist plans for the "improvement of the species" by selective breeding of humans.
>
> We can hope that political perversions of evolutionary theory have been so discredited that they will not happen again. The critical point, however, is that sociobiology is a discipline that attempts to explain why social behavior exists, *not to justify the behavior.*[5]

No. The first critical point is that the "some persons" who were misusing and misapplying scientific ideas were the scientists themselves, and we are only able to see that in hindsight. The second critical point is that *hoping* they will not happen again is a vain endeavor, unless scientists are better educated about the humanistic aspects of what they say and do. And the third critical point is that explaining human behavior by recourse to nature *is* a justification for it, for it implies that the behavior is natural, and that anything different would be to upset the designs of biology. We evolved to walk; hopping is an unnatural form of locomotion to us. To say we evolved to be polygynous and to deny that that is a value judgment is absurdly naive.

A "scientific" idea has ideological power, and when false or antiquated ideas are promoted with the same scientific vigor as valid ones, the activity of science is compromised. The failure to think through the implications of ideas is no longer tolerable in modern human science. It is rather poor scholarship in the first place; but more importantly, scientific ideas can affect people's lives, and scientists are therefore responsible to those people. Scientists did not carry out the Holocaust, but the scientists who held that (for example) Jews were constitutionally inferior and posed a genetic threat to humanity, were more numerous and more vocal in the 1920s than those who did not. That their ideas and arguments inspired—or at least lent a scientific justification to—those who carried it out, is not to the credit of science.

Thus, to ignore the ideological and social value in contemporary statements about human biology is to miss the lesson of history, and to ignore the importance of science in modern life. Conversely, to appreciate it involves immersion in the "humanities"—an awareness of history and culture—which is not inappropriate to the study of humans.

2. TYPOLOGY

Neither the human species, nor any large segment of it, is known to be genetically highly homogeneous. Consequently the prospect of representing it adequately by either a single phenotype or genotype is vanishingly small. Yet one of the most scientific-sounding statements we can make involves generalizing about a group of people, if not about the entire species. It is, of course, fair to generalize, but generalizing about a particular characteristic being human nature implies that the *opposite* characteristic is *not* human nature—and that its bearers either are not human or have a fundamentally different nature.

Biological generalizations apply to a specific reference group, in contrast with another. As noted, for example, there is a characteristically human mode of locomotion, contrasted against those of the great apes. Though we lack any knowledge whatsoever of its genetic basis, it characterizes all normal humans: bipedalism is human nature. Behavioral attributes, however, are far more varied. History tells us that the different natures of groups of people are fantastically malleable through time and across space, which implies that most differences among groups of people are not constitutionally, but socially, rooted. Human nature appears to be extremely diverse, as do the phenotypes of human beings. Representing a group of humans by an idealized member is therefore inadequate both phenotypically and genetically. If humans cannot be accurately captured by a single specimen either genetically or anatomically, it seems unreasonable to expect human *nature* to be so monomorphic either.

3. VARIATION AS PATHOLOGY

Generalizing about human nature implies that deviation from it is *non*-human nature. It is basically a statement about normality, and by implication, a statement about pathology. The more narrowly human nature is defined, the larger the number of people thereby dehumanized—or at least, denormalized.

Locating standards of behavior to the human constitution is one of the classic manners of degrading otherwise human groups. Behavioral variation in the human species is attributable for the most part to cultural history and to individual life experiences. To the extent that there may be genetic variation for behavior, specific behaviors cannot be linked to specific genotypes with any degree of certainty.

There appears, rather, to be a broad spectrum of genotypes resulting in behavioral normality. After all, people from all races, indeed all populations, behave quite normally. Most of the behavior of people classified as deviant is normal as well; for deviancy is often a label earned on the basis of a small sample of one's behavior.[6] Behavioral deviancy thus appears to encompass a broad genetic spectrum as well. And yet the assumption of the Human Genome Project, as it was originally conceived and promoted, involved a goal of sequencing a single haploid stretch of DNA to represent the species. No single stretch could: Not only is there extensive genetic diversity in human populations, but phenotypes, their manifestations, are physiological consequences of the interaction of *two* sets of DNAs—diploidy.

Crime is not like cystic fibrosis, the malfunction of a genetic instruction in a small percentage of people. Neither is promiscuity or poverty or performance in school. How do we know that? Because times change, and the distant offspring of prudes are libertines; those of geniuses are dullards; and those of moguls are waifs. All are common in human societies, and none is more normal than any other. The fallacy lies in defining normality narrowly, usually in terms of specific cultural values; the revision is to recognize the malleability of human behavior on the one hand, and the breadth of the human gene pool, and of human experience itself, on the other.

The flip side of this argument is that pathology can also be taken for normality. When we see genes resulting in diseases whose phenotypes include abnormal behavior, we know genes can affect behavior. But we do not therefore know that subtle variation in normal behavior is the result of subtle variation in the structure of the genes.

4. RACISM

Given the difficulty in narrowly limiting the nature of human groups to a particular word or phrase, it becomes biologically impossible to define the constitution of individual humans by reference to their group membership. Group membership is certainly an important aspect of human social existence, but it has very little in the way of biological meaning.

Instead, the differences between human social groups tend for the most part—certainly the behavioral ones—to be the result of cultural history and life experiences. As far as individual potentialities are concerned, those are discernible only at the level of the individual. To the extent that genotypes set potentials, genotypes are unique and individual. Potentials can be inferred, however, only insofar as they are manifest in performances.

Individual excellence in all human endeavors appears in all groups. Though we can't know for sure, there is no reason to think that potentials vary greatly from population to population. The efficient modern societies will be the ones that cultivate excellence at the individual level, rather than by grouping and judging people by aspects of their ancestry.

And though we may regard as passé the hyper-materialism of equating brain size and intelligence in humans, it remains one of those arguments resurrected every generation by iconoclasts. A recent exchange in the journal *Nature*, for example, centered on the claims of differences in brain size between men and women, and across three races established by self-identification.[7]

J. Philippe Rushton calculated, on the basis of crude skull measurements of army inductees, that the average brain size of Asian males was 1403, of whites 1361, and of blacks 1346 cubic centimeters. This was found even after making adjustments for differences in overall body size. Though it is interesting to note that Asians came out ahead of whites (in contrast to earlier studies, and in harmony with contemporary prejudices about intellect), the difference between whites and blacks is quite small, particularly in relation to total variation in skull size. In addition, men had larger brains than women, even when differences in body size were considered.

Have we thus finally discovered the biological basis for the differences in intelligence that previous generations have always assumed were there? Assuming that the groups defined are natural units, that brain size has been reliably determined, that intelligence has been reliably determined, that brain size is a reasonable estimate of individual intelligence, and that the accomplishments of groups are a straightforward consequence of the intelligences of the individuals composing them—we might well now understand why European and Asian men run the planet.

Ultimately, however, the scientific issues and assumptions are as false as they have always been. First, we must admire the apparent cranial expansion of Asians over the last half-century, when researchers consistently reported their having *smaller* brains than whites.[8] Obviously this implies the possibility of a comparable expansion in blacks. More likely, it implies the possibility of scientists finding just what they expect when the social and political stakes are high.[9]

Second, basic scientific protocol requires that all relevant variables be controlled before drawing conclusions about the cause of an observed difference between samples. But in this case we do not even know what those variables are, or what the appropriate statistical corrections (for example, for body size) may be. Brain size correlates, for example, with age and with nutritional state in early life.[10]

Third, the assumed relationship between brain size and intelligence has always been difficult to establish. Earlier generations correlated brain size with group achievements, to establish a biological basis for the inferiority of the inferior races. This caused some difficulty at the turn of the century, when these scientists found themselves at odds with feminists, whose own inferiority could be established by the same argument from women's lack of achievements.[11] Now individual intelligence is measured by an IQ test, and the correlations that have been reported between IQ and brain size are extremely modest at best.[12]

And finally, the growth, structure, and function of the brain is poorly known, but well enough known to reject the assumption that a measurement of brain *size* yields a precise estimator of brain *quality*. The smallest non-pathological modern human skulls are actually close in cranial capacity to the largest gorilla skulls. But there is no reason to think that the small human brain produces anything but normal human thoughts, and no reason to think that the large gorilla brain produces anything but normal gorilla thoughts.[13] For all we know, the sole advantage to a human of having a big skull over a small skull would come in a head-butting contest. And even then, the thick-vaulted *Homo erectus* would beat out all *Homo sapiens* competition.

After *Nature* ran a negative editorial about the Rushton work, it was slapped with the hackneyed label of "political correctness." But the work in question was, of course, strikingly political in nature—unlike the objective measurements one can take on the brains of sheep or rats, and the statements one would wish to make about their respective intelligence. And the politics of this research is classic reactionary conservativism: establishing that the injustices endured by these historically "disempowered" groups are natural, rooted in their own constitutions.

Interestingly, though there was agreement that women have smaller average brains than men (assuming their brains don't grow in subsequent studies!), they apparently do not have lower average IQs. This obviously would undermine the strict determination of intelligence by brain size, which should already be common sense. (One is certainly hard-pressed to argue that Neanderthals were smarter than modern humans, though their brains were indeed slightly larger.) Thus the relationship between brain *size* and brain *composition* also enters into the comparison. And though women are measured to have smaller brains than men, they seem to have the same average number of neurons in their cerebral cortex, the result of higher concentrations of neurons. The physical basis of intelligence, of course, is quite nebulous regardless of neuronal equality.

By now, this approach to the determination of the average intellectual abilities of group members has degenerated into sophistry. The populations within each "race" vary widely in measured cranial capacity, with the four largest sets of skulls deriving from the aboriginal males of Hawaii, Tierra del Fuego, France, and South Africa, respectively.[14] Even if the average brain *size* differences were real, would there be there any inferences to be drawn about the average *function* of their brains? And what are the implications for evaluating individuals, given that these average differences, both measured and inferred, are relatively small in relation to the overall distribution of the measured variables?

Finally, and most importantly, shouldn't we be arguing about the best ways to ensure that people receive equal opportunities to develop their diverse talents, rather than about the average number or density of neurons encased by a black woman's skull? Whether it is more out of ignorance or malice, directing attention to the brains of human groups today cannot be plausibly considered as anything other than a manner to avoid dealing *socially* with the social issues we face. After all, the study of "the brain" is quite different from the study of "the brains of black people": once neurobiologists break their sample up into such groups, they have crossed the line dividing scientific research that might conceivably have been a dispassionate quest for knowledge, from that which is social, political, and oppressive—and thereby requires far more thorough scrutiny and validation.

Providing explanations for social inequalities as being rooted in nature is a classic pseudoscientific occupation. It has always been welcome, for it provides those in power with a natural validation of their social status. This was as true at the turn of the twentieth century as it is at the turn of the twenty-first—the groups change as the social issues evolve, but the arguments remain eerily unaltered. Ultimately, the assumptions are so large, the data so ambiguous, and the conclusions so overstated, that each generation is forced to reject them. And that generation can always look back in bewilderment at the naive, though similar, suggestions put forward by the previous generation. History is strongly on the side of those who doubt the invocation of nature to explain human social differences.

5. THE CONFUSION OF BIOLOGICAL
AND CULTURAL CATEGORIES

Human races, whatever they are conceived to be, cannot be objectively delimited in space; to a large extent membership within them is culturally rather than biologically defined. We don't know how many there are, where to draw the boundaries between them, or what those boundaries and the people or places they enclose would represent.

Human biological variation, instead, is gradual and continuous. Populations from different parts of the world are obviously often distinguishable from one another as they represent end points of a continuum. Given that race has little biological meaning for humans, we are left with populations.

Human populations differ, and often differ from one another on the average in some particular biological characteristics. In some cases the differences are constitutional and genetic; in others they are not. The social differences that often exist between human groups have served to exaggerate the biological differences. Racial problems in America are mostly social problems: if the social differences among races were minimized, the perceived biological differences would be minimized as well. In the early part of this century, ethnic differences among European immigrant groups to America—such as Irish, Italians, and Jews— and between immigrants and Anglo-Saxon Americans—were widely speculated to be biological in nature. As the economic and social differences among European-Americans diminished, ethnicity remained, but biological constitutional differences among them are no longer widely considered important, if real.

Races are not objective or biological categories. Populations are different from one another, but races are supposed to be large chunks of humanity, and apparently our species doesn't come biologically packaged that way, despite the fact that generations of Euro-Americans have assumed so. At best it comes in lots of small bio-packages. Some earlier writers on the subject assumed there to be three "European" races. In 1939, Carleton Coon established among Europeans a dazzling array of racial diversity: Brunn, Borreby, Alpine, Ladogan, Lappish, Mediterranean, Nordic, Dinaric, Armenoid, and Noric.

Instead today we tend to lump "Europeans" into a single race, because they thereby provide a convenient contrast to "Africans." Though intermarriage has certainly occurred within these European groups, they still retain their identities—but the economic and social parity that they have attained undercuts further regarding their differences as racial (i.e., biological). Once we appreciate that economic and social changes can affect the ostensibly objective scientific perception of racial differences, we can make a projection for the future. Although black people and white people neither can be, nor want to be, *identical* to each other (like the Irish and the Italians, or in Coon's terminology, the Brunn strain and the Dinaricized Mediterranean), the perceived differences between them will also be strongly responsive to economic and social changes.

The categories we acknowledge as races are marked by any number of differences, but the biological differences between them are minimal, reinforced by social and cultural differences. Cultural problems require cultural solutions: when economic parity is attained, the differences between black and white can be expected to follow those between Irish and Italian—distinctions once thought profound and still often identifiable by looks, but ultimately minor.

6. THE EFFECT OF BIOLOGICAL VARIATION ON CULTURAL FORMS

Though cultural distinctions reinforce biological distinctions in maintaining group identity, they also act to differentiate populations biologically from one another. This recognition is a reversal of one of the most popular perceptions of their relationship. We have no good evidence of innate biological differences (i.e., properties of the gene pool) causing cultural differences, but many examples of how cultural differences can cause biological differences.

In particular, resistance to genetic disease may be strongly selected in human populations by virtue of the cultural factors that promote the spread of pathogens. The manner by which this occurs is poorly known, but the widespread epidemics that often follow contact between biologically different populations are an ample testament to the adaptation of gene pools to pathogens, and the lack of such adaptations. Adaptation, ultimately, was the result of long-term coexistence between human communities and the specific diseases they harbored.

Cultural developments are, by contrast, not related to the gene pool in any obvious manner. Earlier generations were able to share credit with their collateral ancestors for having made technological progress in advance of the ancestors of other populations. An appreciation of the social nature of cultural change no longer permits this; innovation is a small part of culture change, and is itself highly dependent upon context. We find creative, innovative people everywhere. Unlike the theorists of earlier generations, we now perceive that our social future lies with identifying talented individuals and developing them, not with assuming the innately superior or inferior abilities of large groups of people, based on the achievements of their ancestors or their cultures.

7. HEREDITARIANISM

Is it nature or nurture? Heredity or environment? Hard science or soft science?

The development of genetics in the 20th century made the "nature" position sound more credible, but of course the hereditarian philosophy—the assumption that "blood will tell," that biological inheritance is one's destiny—long pre-dates the rise of the science of genetics.

Hereditarianism piggybacks on genetics. By talking about specific named behaviors in contemporary genetic terms, we make the old hereditarian point of view sound more valid. And yet the old arguments and assumptions continue to arise each generation, as if no attention had been paid to the advances made in the study of heredity. The anthropologist Kroeber pointed out in 1916 that the eugenics movement, though cloaked in contemporary science, was fundamentally just a folk theory of heredity that didn't incorporate modern scientific understandings at all, but simply reinforced old cultural values and was persuasive on that basis.

Today new genetic advances occur weekly. Still, however, they enter into a set of cultural values that we and the scientists bring to the work to interpret it. History shows that the assumption that behavioral differences are at root innate is one of the most abused of scientific conclusions. It is a powerful source of validation for the status quo, for passing on responsibility for social problems, and for unfairly ostracizing people in large numbers. Consequently these must be the among the most carefully scrutinized of scientific conclusions.

The great advances in genetics in this century have not brought about great advances in our understanding of why crime exists, the nature of variation in moral standards, or the association of specific behaviors with specific populations at specific times. Rather, the arguments on behalf of the hereditarian position tend to rest on the same kinds of arguments and data they have always rested on: anecdotes, twins, skulls, and prejudices.

Genetic research is of obvious value, but the answers it provides are to medical problems, not social problems. The social solutions involve acknowledging a joint responsibility as a prelude to action. We are part of the problem, for we are part of society. Unfortunately, most people don't like to admit their responsibility; others make better targets. It has always been easier to blame their gene pools than to begin reforming the circumstances in which they exist, and the values they have been transmitted. It is easier to think about solving social problems in the third person than in the first.

That is also why students of ecology decry population explosions, yet still themselves have babies. Environmentally active families recycle their newspapers, yet nevertheless maintain three cars and eschew public transportation. The trouble with devising and implementing social solutions to social problems is that they require first-person sacrifice, a strong sense of responsibility, and the wide distribution of that responsibility.

The study of human genetic variation is among the most rewarding of scientific endeavors. Understanding the ways in which the differences among people are patterned helps us to understand our own nature and existence more fully. It also helps us to see the ways in which science is not what we have popularly imagined it to be. Not so much the objective collection of facts, science is carried out by people with their own backgrounds and values. Often it has proven difficult to distinguish competent from incompetent conclusions, when they tend to bear so strongly on our conception of who we are, and how we relate to others.

Ultimately, if we learn little about variation in human behavior from genetics, that little is itself a great deal. It tells us where we should be looking, by showing us where answers have not been found. The study of human genetics, then, tells us much about human genetics; that it tells us little about other phenomena is not a considerable problem. One does not find fault with a refrigerator for not cooling the room: that is a job for other equipment. The prematurity of Mendel's work, it is now recognized, stems from the fact that Mendel worked exclusively on principles of genetic transmission at a time when heredity subsumed both transmission and ontogenetic growth. When those two processes were disentangled in the late 19th century, Mendel's work exclusively on one of them became interpretable.[15] Likewise, the cultural and biological processes that affect our species must be conceptually disentangled before we can make true intellectual progress.

No one is able to step out of time and culture, and see how their own ideas are shaped not by data, but by ideology. That only works in retrospect. Undoubtedly, students a generation hence will find the flaws with our science as we do with our own intellectual predecessors. Unlike culture as a whole, its subset science *does* progress, and it does so by explicitly identifying and transcending the mistakes of earlier generations.

But it is tempting to commit those mistakes again and again.

NOTES

1. Koshland (1990:189).
2. Koshland (1989:189).
3. Jennings (1925:25–26).
4. Proctor (1991).
5. Alcock (1993:545–46, emphasis in original).
6. Michel Foucault (1984), in particular, has called attention to the transfer

in recent European history of the adjective "homosexual" from referring to an act to referring to a person.

7. Maddox (1992), Becker (1992), Rushton (1992a), Ankney (1992a), Schluter (1992), Lynn (1992), Tsai (1992), Blest (1992), Brand (1992), Van Valen (1992), Jones (1992), Rushton (1992b), Ankney (1992b).

8. See, for example, Gates (1946).

9. Gould (1978).

10. Tobias (1970).

11. Soloway (1990).

12. Van Valen (1974).

13. Dart (1956), Tobias (1970).

14. Groves (1990).

15. Bowler (1991).

Appendix

DNA Structure and Function

One of the greatest triumphs of 20th-century biology has been to elaborate the biochemical structure underlying genetic inheritance. DNA (deoxyribonucleic acid) is the molecule that the genes are composed of, and that is physically passed down from parent to child. Its structure is the famous "double helix," for which James Watson and Francis Crick won the Nobel Prize.

The molecule can be conceived most simply as a ladder, consisting fundamentally of two opposing struts and many rungs. The "struts" are composed of alternating sugar and phosphate molecules, the sugar being the part to which the "rungs" are attached. The rungs are a series of bases, whose particular sequence literally composes the genetic information.

Each strand of the double-stranded backbone is built up of a derivative of the sugar ribose, whose five carbon atoms are numbered by convention, as shown in Figure A.1. Carbon number 2 (called 2', or "two-prime," to distinguish it from the number 2 carbon of one of the bases in the DNA molecule) lacks an oxygen atom possessed by its neighboring 3'-carbon. If the 2'-carbon had this oxygen, it would be the sugar ribose; lacking it, the sugar is deoxyribose.

Attached to the 3'-carbon is a phosphate group, derived from phosphoric acid, the "acid" in DNA, which in turn is linked to the 5'-carbon of the sugar below it. The 1'-carbon is the attachment point for the base (Figure A.2).

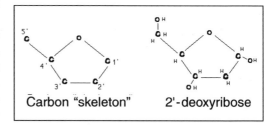

Carbon "skeleton"   2'-deoxyribose

Figure A.1. The sugar component of DNA.
C represents carbon, O represents oxygen, H represent hydrogen.

Each side of the backbone is thus a long series of alternating sugars and phosphates, the first sugar having an unattached 5'-carbon, and the last having an unattached 3'-carbon. It can thus be considered to have an element of directionality, running from 5' to 3'.

The other DNA stand is composed the same way, but in the opposite orientation (Figure A.3). Where the first sugar on one strand has an unattached 5'-carbon, the other DNA strand has an unattached 3'-carbon opposite it. This strand can thus be considered to run from 3' to 5'. In other words, the strands are not simply mirror images; they are polar opposites.

It is not, however, the sugar-phosphate backbone that bears the genetic information; it is the sequence of bases within the DNA chain.

The bases adenine (A) and guanine (G) are larger and are called purines; cytosine (C) and thymine (T) are smaller and are called pyrimidines. (Alternatively, the word *nucleotide* can be used for *base*; technically it refers to a base attached to a sugar and a phosphate.) The insight of Watson and Crick lay in inferring that a purine attached to a sugar on one strand would invariably be found with a particular pyrimidine across from it, attached to the ribose on the other strand. That specificity was determined by the spatial structure of the molecules, and dictated that adenine would be paired with thymine, and cytosine with guanine. The pairs of bases are held together by sharing hydrogen atoms: the A-T base pair by two, and the G-C base pair by three. This highly specific pairing and bonding is as fundamental an aspect of the life processes as any yet discovered.

We can thus draw the DNA molecule schematically as shown in Figure A.4, with two parallel sugar-phosphate chains running in opposite orientations, and a linear array of bases down the middle, with a series

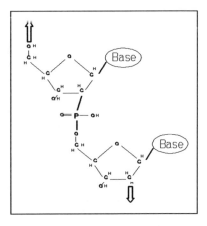

Figure A.2.   Structure of the sugar-phosphate linkage in DNA. P represents phosphorus.

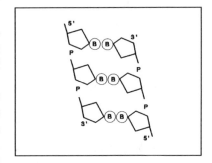

Figure A.3.   Orientation of the opposing strands of DNA.

of complementary bases opposing them. Since the bases are complementary to one another, one need only know the base sequence of one strand to generate the other. Further, since the genetic information is contained within the specificity of the sequence of bases, one can ignore the sugar-phosphate backbone for the purposes of discussing DNA sequences.

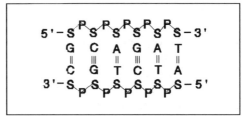

Figure A.4. DNA as a sequence of bases on one strand, with a complementary sequence on the other.

Thus, the DNA sequence given in Figure A.4 could be reduced to GCAGAT, with an implicit understanding that there is a complementary strand reading CGTCTA, and that both strands contain alternating sugars and phosphates, in opposite orientations on the two strands. Lengths of DNA can be given, therefore, in units of base-pairs (bp) or kilobases (kb).

By the word *gene* we generally mean a functional segment of DNA. Most DNA, as noted in Chapter 8, is not in fact functional. Thus, to a large extent genes are simply rare bits of DNA that are significant by virtue of having a function.

The function of a gene is determined by the way a cell is able to implement the information the gene encodes. Genes, after all, are generally passive structures—their *information* is what the cell needs. That information is extracted by the creation of a transient intermediary molecule called RNA, whose own structure contains a copy of the gene. Thus, genes act by virtue of having RNA transcribed from their DNA sequence—a messenger or mRNA, similar in structure to DNA—which then travels out of the cell nucleus and has its nucleotide sequence translated into a specific protein.

The structure of a

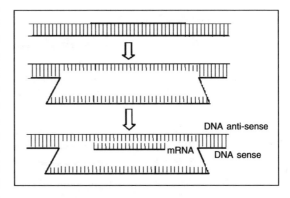

Figure A.5. RNA is transcribed from one strand of DNA; it is thus similar in base sequence to one strand, and complementary in sequence to the other.

coding gene is defined in relation to the fate of its mRNA transcript Since only one strand contains the biological information, an mRNA molecule is polymerized on only one strand (Figure A.5). It is thus complementary to that strand, and therefore identical to the other strand. This defines one DNA strand, the one similar to the RNA sequence, as the *sense* strand; and the other strand, complementary to the RNA, as the *anti-sense* strand. Published DNA sequences are those of the sense strand.

The efficiency of transcription is strongly affected by small groups of bases before the beginning of the gene, which are known as promoters. A primary transcript, or precursor mRNA, is processed in three ways, which define regions of the DNA from which it is derived. At the beginning of the transcript, the first nucleotide is modified, or "capped," which defines the "cap" site of the DNA (Figure A.6). Near the end of the transcript, the mRNA is cleaved shortly beyond the characteristic base sequence AATAAA, and a long chain of adenines is added, thus defining the poly-adenylation site. And within the mRNA itself, some regions are deleted (introns), and the segments that remain are spliced together (exons). The result is a contiguous coding sequence flanked by untranslated RNA regions. The protein product is thus encoded by a relatively small portion of what we designate as a gene.

The end products of some genes, however, are not proteins, but sim-

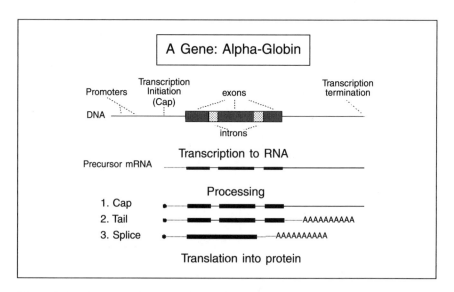

Figure A.6. Gene structure in relation to mRNA processing. The RNA transcript is far longer than its actual coding sequence, and initially far longer still. Thus, much of a gene is itself "non-coding."

ply functional RNAs. This means that "coding for protein" is too strict a definition for a gene.

Some stretches of DNA do not themselves code for proteins and are not even transcribed into RNA, but affect the transcriptional efficiency of a nearby gene. Others serve almost entirely passive functions—they can serve as "binding sites" for the products of other genes, or can perhaps affect the higher-dimensional structure of the DNA molecule, as it bends and folds into the relatively large structure known as a chromosome. Thus the functional units of DNA may be far broader to define than simply those which code for active proteins.

References

Alcock, J. (1993). *Animal Behavior.* Sunderland, MA: Sinauer.

Allen, W., and Ostrer, H. (1993). "Anticipating Unfair Uses of Genetic Information." *American Journal of Human Genetics* 53:16–21.

Alper, J. S., and Natowicz, M. R. (1993). "Genetic Discrimination and the Public Entities and Public Titles of the Americans with Disabilities Act." *American Journal of Human Genetics,* 53:26–32.

Alpers, M. (1992). "Kuru." Pp. 313–34 in *Human Biology in Papua New Guinea: The Small Cosmos,* edited by R. D. Attenborough and M. P. Alpers. New York: Oxford University Press.

Anderson, A. M. (1982). "The Great Japanese IQ Increase." *Nature* 297:190–81.

Andrews, P. (1992). "Evolution and Environment in the Hominoidea." *Nature,* 360:641–46.

Ankney, C. D. (1992a). "Differences in Brain Size." *Nature* 358:532.

Ankney, C. D. (1992b). "The Brain Size/IQ Debate." *Nature* 360:292.

Anonymous (1930). "Three Notable Books on Human Inheritance." *Journal of Heredity* 21:171–72.

Anonymous (1992a). "Reporting of Race and Ethnicity in the National Notifiable Diseases Surveillance System, 1990." *MMWR* 41:653–57.

Anonymous (1992b). "Death Rates for Minority Infants Were Underestimated, Study Says." *New York Times,* January 8.

Ardrey, R. (1961). *African Genesis.* New York: Atheneum.

Ardrey, R. (1966). *The Territorial Imperative.* New York: Atheneum.

Ashe, A. R., Jr. (1993a). *A Hard Road to Glory: The African-American Athlete in Basketball.* New York: Amistad.

Ashe, A. R., Jr. (1993b). *A Hard Road to Glory: The African-American Athlete in Track and Field.* New York: Amistad.

Ayala, F. (1969). "Evolution of Fitness. V. Rate of Evolution in Irradiated Populations of Drosophila." *Proceedings of the National Academy of Sciences* 63:790–93.

Bahn, P., and Vertut, J. (1988). *Images of the Ice Age.* New York: Facts on File.

Bailey, M., and Pillard, R. (1991). "Are Some People Born Gay?" *New York Times,* December 7.

Baker, P. T., and Little, M. A., eds. (1976). *Man in the Andes.* Stroudsburg, PA: Dowden, Hutchinson and Ross.

Bank, A. (1985). "Genetic Defects in the Thalassemias." *Current Topics in Hematology* 5:1–23.

Banton, M. (1987). *Racial Theories.* New York: Cambridge University Press.

Dunton, M, and Harwood J (1975), *The Race Concept*. London: David and Charles.

Barash, D. (1979). *The Whisperings Within*. New York: Penguin.

Barash, D. (1993). "Sex, Violence, and Sociobiology." *Science* 262: (no page number).

Barinaga, M. (1989). "Manic Depression Gene Put in Limbo." *Science* 245:886–87.

Barinaga, M. (1991). "How Long Is the Human Life-Span?" *Science* 254:936–38.

Barkan, E. A. (1992). *The Retreat of Scientific Racism*. New York: Cambridge University Press.

Barker, D. J. P. (1989). "Rise and Fall of Western Diseases." *Nature* 338:371–72.

Barnett, H. G. (1953). *Innovation: The Basis of Cultural Change*. New York: McGraw-Hill.

Barzun, J. ([1937] 1965). *Race: A Study in Superstition*. New York: Harper and Row.

Basalla, G. (1988). *The Evolution of Technology*. New York: Cambridge University Press.

Bateman, A. J. (1948). "Intra-Sexual Selection in Drosophila." *Heredity* 2:349–68.

Bateson, W. ([1905] 1928). "Evolution for Amateurs." Pp. 449–55 in *William Bateson, Naturalist*, edited by B. Bateson. Cambridge: Cambridge University Press

Bateson, W. ([1919] 1928). "Common-Sense in Racial Problems." Pp. 371–87 in *William Bateson, Naturalist*, edited by B. Bateson. Cambridge: Cambridge University Press.

Batzer, M. A., Gudi, V. A., Mena, J. C., Foltz, D. W., Herrera, R. J., and Deininger, P. L. (1992). "Amplification Dynamics of Human-Specific (HS) Alu Family Members." *Nucleic Acids Research* 19:3619–23.

Baur, E., Fischer, E., and Lenz, F. (1931). *Human Heredity*, translated by Eden and Cedar Paul. New York: Macmillan.

Beatty, J. (1987). "Weighing the Risks: Stalemate in the Classical/Balance Controversy." *Journal of the History of Biology* 20:289–319.

Beatty, J. (1994). "The Proximate/Ultimate Distinction in the Multiple Careers of Ernst Mayr." *Biology and Philosophy* 9:333–56.

Becker, B. A. (1992). "Differences in Brain Size." *Nature* 358:532.

Beckwith, J., Geller, L., and Sarkar, S. (1991). "IQ and Heredity." *Science* 252:191.

Begley, S., Murr, A., Springen, K., Gordon, J., and Harrison, J. (1987). "All About Twins." *Newsweek* (November 23):58–69.

Bell, J. I., Todd, J. A., and McDevitt, H. O. (1989). "The Molecular Basis of HLA-Disease Association." pp. 1–41 in *Advances in Human Genetics, Vol. 18*, edited by H. Harris and K. Hirschhorn. New York: Plenum.

Bendyshe, T. (1865). "The History of Anthropology." *Memoirs of the Anthropological Society of London* 1:335–458.

Bennett, L., Jr. ([1962] 1966). *Before the Mayflower: A History of the Negro in America, 1619–1964*. Baltimore, MD: Penguin.

Bernal, J. D. ([1939] 1967). *The Social Function of Science*. Cambridge, MA: MIT Press.

Berry, R. J. (1968). "The Biology of Non-Metrical Variation in Mice and Men." Pp. 103-134 in *The Skeletal Biology of Earlier Human Populations*, edited by D. R. Brothwell. Oxford: Pergamon.

Betzig, L., Borgerhoff Mulder, M., and Turke, P., eds. (1988). *Human Reproductive Behaviour: A Darwinian Perspective.* Cambridge: Cambridge University Press.

Beutler, E. (1992). "Gaucher Disease: New Molecular Approaches to Diagnosis and Treatment." *Science* 256:794–98.

Biasutti, R. (1958). *Le Razze e i Popoli della Terra,* 3d edition. Turin: Unione Tipigrafico–Editrice Torinese.

Biddiss, M. D. (1970). *Father of Racist Ideology: The Social and Political Thought of Count Gobineau.* London: Weidenfeld and Nicolson.

Billings, P. R., Kohn, M. A., de Cuevas, M., Beckwith, J., Alper, J. S., and Natowicz, M. R. (1992). "Discrimination as a Consequence of Genetic Testing." *American Journal of Human Genetics* 50:476–82.

Bilsborough, A. (1992). *Human Evolution.* London: Blackie.

Birdsell, J. B. (1950). "Some Implications of the Genetical Concept of Race in Terms of Spatial Analysis." *Cold Spring Harbor Symposia on Quantitative Biology* 15:259–314.

Birdsell, J. B. (1952). "On Various Levels of Objectivity in Genetical Anthropology." *American Journal of Physical Anthropology* 10:355–62.

Birdsell, J. B. (1963). "Review of 'The Origin of Races.'" *Quarterly Review of Biology* 38:178–85.

Birdsell, J. B. (1987). "Some Reflections on Fifty Years in Biological Anthropology." *Annual Review of Anthropology* 16:1–12.

Black, F. L. (1991). "Reasons for the Failure of Genetic Classifications of South Amerind Populations." *Human Biology* 63:763–74.

Black, F. L. (1992). "Why Did They Die?" *Science* 258:1739–40.

Blest, A. D. (1992). "Brain Size Differences." *Nature* 359:182.

Bloom, B. R., and Murray, C. J. L. (1992). "Tuberculosis: Commentary on a Reemergent Killer." *Science* 257:1055–64.

Blumenbach, J. F. ([1795] 1865). *On the Natural Variety of Mankind,* 3d edition. In *The Anthropological Treatises of Johann Friedrich Blumenbach,* translated and edited by Thomas Bendyshe. London: Longman, Green.

Boas, F. (1896). "The Limitations of the Comparative Method in Anthropology." *Science* 4:901–8.

Boas, F. (1911). *The Mind of Primitive Man.* New York: Macmillan.

Boas, F. (1912). "Changes in the Bodily Form of Descendants of Immigrants." *American Anthropologist* 14:530–62.

Boas, F. (1916). "Eugenics." *Scientific Monthly* 3:471–478.

Boas, F. (1924). "The Question of Racial Purity." *The American Mercury* 3:163–69.

Boas, F. (1928). *Anthropology and Modern Life.* New York: Norton.

Boaz, N. T. (1988). "Status of Australopithecus Afarensis." *Yearbook of Physical Anthropology* 31:85–113.

Bodmer, W. F. (1986). "Human Genetics: The Molecular Challenge." *Cold Spring Harbor Symposium on Molecular Biology* 51:1–13.

Bogin, B. (1988). "Rural-to-Urban Migration." Pp. 90–129 in *Biological Aspects of Human Migration,* edited by C. G. N. Mascie-Taylor and G. W. Lasker. New York: Cambridge University Press.

Bogin, B. (1988). *Patterns of Human Growth.* New York: Cambridge University Press.

Bonner, J. T. (1902). *The Evolution of Culture in Animals*. Princeton, NJ: Princeton University Press.

Bouchard, T. T. j. Jr., Lykken, D. T., McGue, M., Segal, N. L., and Tellegan, A. (1990). "Sources of Human Psychological Differences: The Minnesota Study of Twins Reared Apart." *Science* 250:223–28.

Bowcock, A. M., Kidd, J. R., Mountain, J. L., Hebert, J. M., Carotenuto, L., Kidd, K. K., and Cavalli-Sforza, L. L. (1991). "Drift, Admixture, and Selection in Human Evolution: A Study with DNA Polymorphisms." *Proceedings of the National Academy of Sciences* 88:839–43.

Bowcock, A. M., Ruiz-Linares, A., Tomfohrde, J., Minch, E., Kidd, J. R., and Cavalli-Sforza, L. L. (1994). "High Resolution of Human Evolutionary Trees with Polymorphic Microsatellites." *Nature* 368:455–57.

Bowler, P. J. (1973a). "Bonnet and Buffon: Theories of Generation and the Problem of Species." *Journal of the History of Biology* 6:259–81.

Bowler, P. J. (1973b). *The Eclipse of Darwinism: Anti-Darwinian Evolution Theories i the Decades Around 1900*. Baltimore: Johns Hopkins University Press.

Bowler, P. J. (1983). *The Eclipse of Darwinism*. Baltimore: Johns Hopkins University Press.

Bowler, P. J. (1984). *Evolution: The History of an Idea*. Berkeley: University of California Press.

Bowler, P. J. (1980). *The Mendelian Revolution*. Baltimore: Johns Hopkins University Press.

Box, J. F. (1978). *R. A. Fisher: The Life of a Scientist*. New York: Wiley.

Boyd, R., and Richerson, P. (1985). *Culture and the Evolutionary Process*. Chicago: University of Chicago Press.

Boyd, W. C. (1947). "The Use of Genetically Determined Characters, Especially Serological Factors Such as Rh, in Physical Anthropology." *Southwestern Journal of Anthropology* 3:32–49.

Boyd, W. C. (1949). "Systematics, Evolution, and Anthropology in the Light of Immunology." *Quarterly Review of Biology* 24:102–8.

Boyd, W. C. (1950). *Genetics and the Races of Man*. Boston: Little, Brown.

Boyd, W. C. (1963). "Genetics and the Human Race." *Science* 140:1057–65.

Brace, C. L. (1964). "On the Race Concept." *Current Anthropology* 313–20.

Brand, C. R. (1992). "Sizing-Up the Brain." *Nature* 359:768.

Brandt, A. (1988). "The Syphilis Epidemic and Its Relation to AIDS." *Science* 239:375–80.

Brauer, G., and Mbua, E. (1992). "Homo Erectus Features Used in Cladistics and Their Variability in Asian and African Hominids." *Journal of Human Evolution* 22:79–108.

Britten, R. J., and Kohne, D. E. (1968). "Repeated Sequences in DNA." *Science* 161:529–40.

Brooklyn, Museum (1988). *Cleopatra's Egypt: Age of the Ptolemies*. New York: Brooklyn Museum.

Broom, R. (1938). "The Pleistocene Anthropoid Apes of South Africa." *Nature* 142:377–79.

Brothwell, D. (1968). "Introducing the Field." Pp. 1–18 in *The Skeletal Biology of Earlier Human Populations*, edited by D. R. Brothwell. Oxford: Pergamon.

Brunner, H. G., Nelen, M. R., Zandvoort, P. van, Abeling, N. G. G. M., Gennip, A. H. van, Wolters, E. C., Kuiper, M. A., Ropers, H. H., and Oost, B. A. van (1993). "X-Linked Borderline Mental Retardation with Prominent Behavioral Disturbance: Phenotype, Genetic Localization, and Evidence for Disturbed Monoamine Metabolism." *American Journal of Human Genetics* 52:1032–39.

Buffon, Count de (1749–1804). *Histoire Naturelle, Gènèrelle et Particulière.* Paris: Imprimerie Royale, later Plassans.

Buffon, Count de ([1749] 1812). "Varieties of the Human Species." In *Natural History, General and Particular,* translated by William Smellie. London: T. Cadell and W. Davies.

Burkitt, D. P. (1981). "Geography of Disease: Purpose of and Possibilities from Geographical Medicine." Pp. 133–51 in *Biocultural Aspects of Disease,* edited by H. Rothschild. Orlando, FL: Academic Press.

Bury, J. B. (1932). *The Idea of Progress: An Inquiry into Its Origin and Growth.* New York: Macmillan.

Butterfield, H. ([1931] 1965). *The Whig Interpretation of History.* New York: Norton.

Byerley, W. F. (1989). "Genetic Linkage Revisited." *Nature* 340:340–41.

Cann, R. L., Stoneking, M., and Wilson, A. C. (1987). "Mitochondrial DNA and Human Evolution." *Nature* 325:31–36.

Capron, A. M. (1990). "Which Ills to Bear? Reevaluating the 'Threat' of Modern Genetics." *Emory Law Journal* 39:665–96.

Capron, C., and Duyme, M. (1989). "Assessment of Effects of Socio-Economic Status on IQ in a Full Cross-Fostering Study." *Nature* 340:552–53.

Carlin, J. (1989). "The Group." *New Republic* (November 27):21–23.

Caro, T. M., and Laurenson, M. K. (1994). "Ecological and Genetic Factors in Conservation: A Cautionary Tale." *Science* 263:485–86.

Carrier, J. M., and Gellert, G. (1991). "Biology and Homosexuality." *Science* 254:630.

Castle, W. E. (1926). "Biological and Social Consequences of Race Crossing." *American Journal of Physical Anthropology* 9:145–56.

Castle, W. E. (1930a). *Genetics and Eugenics.* Cambridge: Harvard University Press.

Castle, W. E. (1930b). "Race Mixture and Physical Disharmonies." *Science* 71:603–6.

Cavalier-Smith, T., ed. (1985). *The Evolution of Genome Size.* New York: Wiley.

Cavalli-Sforza, L. L. (1991). "Genes, Peoples and Languages." *Scientific American* 265(5):104–10.

Cavalli-Sforza, L. L., and Feldman, M. W. (1981). *Cultural Transmission and Evolution: A Quantitative Approach.* Princeton, NJ: Princeton University Press.

Cavalli-Sforza, L. L., and Edwards, A. W. F. (1964). "Analysis of Human Evolution." Pp. 923–33 in *Genetics Today,* edited by S. J. Geertz. Oxford: Pergamon.

Cavalli-Sforza, L. L., Menozzi, P., and Piazza, A. (1993). "Demic Expansions and Human Evolution." *Science* 259:639–46.

Cavalli-Sforza, L. L., Piazza, A., Menozzi, P., and Mountain, J. (1988). "Reconstruction of Human Evolution: Bringing Together Genetic, Archaeological, and Linguistic Data." *Proceedings of the National Academy of Sciences* 85:6002–6.

Cavalli-Sforza, L. L., Wilson , A. C., Cantor, C. R., Cook-Deegan, R. M., and King, M.-C. (1991). "Call for a Worldwide Survey of Human Genetic Diversity: A Vanishing Opportunity for the Human Genome Project." *Genomics* 11:490–91.

Chakraborty, R., Kamboh, M., Nwankwo, M., and Ferrell, R. E. (1992). "Caucasian Genes in American Blacks: New Data." *American Journal of Human Genetics* 50:145–55.

Chamberlain, H. S. ([1899] 1910). *Foundations of the Nineteenth Century*. New York: Lane.

Chambers, R. (1844). *Vestiges of the Natural History of Creation*. London: J. Churchill [published anonymously].

Chase, A. (1980). *The Legacy of Malthus: The Social Costs of the New Scientific Racism*. Urbana: University of Illinois Press.

Chase, A. (1982). *Magic Shots*. New York: Morrow.

Chase, G. A., and McKusick, V. A. (1972). "Controversy in Human Genetics: Founder Effect in Tay-Sachs Disease." *American Journal of Human Genetics* 24:339–40.

Chetverikov, S. S. ([1926] 1961). "On Certain Aspects of the Evolutionary Process from the Standpoint of Modern Genetics." *Proceedings of the American Philosophical Society* 105:167–95.

Classen, C. (1993). *Worlds of Sense*. New York: Routledge.

Coale, A. J. (1983). "Recent Trends in Fertility in Less Developed Countries." *Science* 221:828–832.

Cobb, W. M. (1936). "Race and Runners." *Journal of Health and Physical Education* 7:1–9.

Cobb, W. M. (1947). "Does Science Favor Negro Athletes?" *Negro Digest* (May):47–48.

Cohen, M. N. (1977). *The Food Crisis in Prehistory*. New Haven, CT: Yale University Press.

Cohen, M. N. (1989). *Health and the Rise of Civilization*. New Haven, CT: Yale University Press.

Cohen, M. N., and Armelagos, G., eds. (1984). *Paleopathology at the Origins of Agriculture*. New York: Academic Press.

Coleman, D. A. (1990). "The Demography of Social Class." Pp. 59–116 in *Biosocial Aspects of Social Class*, edited by C. G. N. Mascie-Taylor. New York: Oxford University Press.

Collins, F. (1992). "Cystic Fibrosis: Molecular Biology and Therapeutic Implications." *Science* 256:774–79.

Collins, F., and Weissman, S. (1984). "The Molecular Genetics of Human Hemoglobin." *Progress in Nucleic Acid Research and Molecular Biology* 31:315–439.

Conklin, E. G. (1922). *Heredity and Environment*, 5th edition. Princeton, NJ: Princeton University Press.

Coon, C. S. (1939). *The Races of Europe*. New York: Macmillan.

Coon, C. S. (1962). *The Origin of Races*. New York: Alfred M. Knopf.

Coon, C. S. (1965). *The Living Races of Man*. New York: Alfred M. Knopf.

Coon, C. S. (1968). "Comment on 'Bogus Science.'" *Journal of Heredity* 59:275.

Coon, C. S. (1981). *Adventures and Discoveries.* Englewood Cliffs, NJ: Prentice-Hall.

Cowan, R. S. (1992). "Genetic Technology and Reproductive Choice: An Ethics for Autonomy." Pp. 244–63 in *The Code of Codes,* edited by D. J. Kevles and L. Hood. Cambridge, MA: Harvard University Press.

Crosby, A. W. (1986). *Ecological Imperialism: The Biological Expansion of Europe, 900–1900.* New York: Cambridge University Press.

Crow, J. F. (1988). "Eighty Years Ago: The Beginnings of Population Genetics." *Genetics* 119:473–76.

Crow, J. F. (1990). "Sewall Wright's Place in Twentieth-Century Biology." *Journal of the History of Biology* 23:57–89.

Culliton, B. (1976). "Penicillin-Resistant Gonorrhea: New Strain Spreading Worldwide." *Science* 194:1395–97.

Curran, J. W., Jaffe, H. W., Hardy, A. M., Morgan, W. M., Selik, R. M., and Dondero, T. J. (1988). "Epidemiology of HIV Infection and AIDS in the United States." *Science* 239:610–16.

Cuvier, G. (1829). *Le Règne Animal.* Paris: Déterville.

Daly, M., and Wilson, M. (1983). *Sex, Evolution, and Behavior.* Boston: PWS.

Darrow, C. (1926). "The Eugenics Cult." *The American Mercury* 8:129–37.

Dart, R. A. (1925). "Australopithecus Africanus: The Man-Ape of South Africa." *Nature* 115:195–99.

Dart, R. A. (1956). "The Relationships of Brain Size and Brain Pattern to Human Status." *South African Journal of Medical Science* 21:23–45.

Darwin, C. (1859). *On the Origin of Species by Means of Natural Selection, or the Preservation of Favoured Races in the Struggle for Life.* London: John Murray.

Davenport, C. B. (1911). *Heredity in Relation to Eugenics.* New York: Henry Holt.

Davenport, C. B. (1917). "The Effects of Race Intermingling." *Proceedings of the American Philosophical Society* 56:364–68.

Davenport, C. B. (1921). "Research in Eugenics." *Science* 54:391–97.

Davenport, C. B., and Steggerda, M. (1929). *Race Crossing in Jamaica.* Washington, DC: Carnegie Institution.

Davies, K. (1993). "Protection and Susceptibility." *Nature* 362:478.

Davis, B. D., et al. (1990). "The Human Genome and Other Initiatives." *Science* 249:342–43.

Dawkins, R. (1986). *The Blind Watchmaker.* New York: Norton.

De Waal, F. B. M. (1989). *Peacemaking among Primates.* Cambridge, MA: Harvard University Press.

Dean, D., and Delson, E. (1992). "Second Gorilla or Third Chimp?" *Nature* 359:676–77.

Dean, G. (1971). *The Poryphyrias,* 2d edition. Philadelphia: Lippincott.

Detera-Wadleigh, S. D., Goldin, L. R., Sherrington, R., Encio, I., de Miguel, C., Berrettini, W., Gurling, H., and Gershon, E. S. (1989). "Exclusion of Linkage to 5q11-13 in Families with Schizophrenia and Other Psychiatric Disorders." *Nature* 340:391–93.

Devor, E. J., and Cloninger, C. R. (1989). "Genetics of Alcoholism." *Annual Review of Genetics* 23:19–36.

Diamond, J. (1992). *The Third Chimpanzee*. New York: Harper Collins.

Diamond, R. (1993). "Genetics and Male Sexual Orientation." *Science* 261:1258–59.

Dickens, B. M. (1988). "Legal Rights and Duties in the AIDS Epidemic." *Science* 239:580–86.

Dixon, R. B. (1923). *The Racial History of Man*. New York: Scribner's.

Dobzhansky, T. (1937). *Genetics and the Origin of Species*. New York: Columbia University Press.

Dobzhansky, T. (1955). "A Review of Some Fundamental Concepts and Problems of Population Genetics." *Cold Spring Harbor Symposium on Quantitative Biology* 20:1–15.

Dobzhansky, T. (1959). "Variation and Evolution." *Proceedings of the American Philosophical Society* 103:252–63.

Dobzhansky, T. (1963). "Evolutionary and Population Genetics." *Science* 142:1131–35.

Dobzhansky, T. (1968). "More Bogus "Science" of Race Prejudice." *Journal of Heredity* 59:102–4.

Dobzhansky, T. (1970). *Genetics of the Evolutionary Process*. New York: Columbia University Press.

Dorfman, D. D. (1978). "The Cyril Burt Question: New Findings." *Science* 201:1177–86.

Dudley, R. M. (1991). "IQ and Heredity." *Science* 252:191.

Dulbecco, R. (1986). "A Turning Point in Cancer Research: Sequencing the Human Genome." *Science* 231:1055–56.

Dunbar, R. (1993). "Behavioural Adaptation." Pp. 73–98 in *Human Adaptation*, edited by G. A. Harrison. New York: Oxford University Press.

Dunn, L. C. (1965). *A Short History of Genetics*. New York: McGraw-Hill.

Durham, W. (1991). *Coevolution*. Stanford, CA: Stanford University Press.

Duster, T. (1990). *Backdoor to Eugenics*. New York: Routledge.

East, E. M. (1927). *Heredity and Human Affairs*. New York: Charles Scribner's.

Eddy, J. H., Jr. (1984). "Buffon, Organic Alterations, and Man." *Studies in the History of Biology* 7:1–45.

Egeland, J. A., Gerhard, D. S., Pauls, D. L., Sussex, J. N., Kidd, K. K., Allen, C. R., Hostetter, A. M., and Housman, D. E. (1987). "Bipolar Affective Disorders Linked to DNA Markers on Chromosome 11." *Nature* 325:783–87.

Ehrenreich, B. (1993). "The Gap between Gay and Straight." *Time* (May 10):76.

Eisensmith, R. C., and Woo, S. L. C. (1991). "Phenylketonuria, Molecular Genetics." Pp. 863–70 in *The Encyclopedia of Human Biology*, Volume 5, edited by R. Dulbecco. San Diego: Academic Press.

Eldredge, N. (1982). "Phenomenological Levels and Evolutionary Rates." *Systematic Zoology* 31:338–47.

Eldredge, N. (1985). *Unfinished Synthesis: Biological Hierarchies and Modern Evolutionary Thought*. New York: Oxford University Press.

Engels, F. ([1840] 1940). *Dialectics of Nature*. New York: International Publishers.

Fausto-Sterling, A., and Balaban, E. (1993). "Genetics and Male Sexual Orientation." *Science* 261:1257.

Fellows, O. E., and Milliken, S. F. (1972). *Buffon*. New York: Twayne.

Ferguson, B. (1990). "Explaining War." Pp. 26–55 in *The Anthropology of War*, edited by J. Haas. New York: Cambridge University Press.

Ferris, S. D., Brown, W. M., Davidson, W. S., and Wilson, A. C. (1981). "Extensive Polymorphism in the Mitochondrial DNA of Apes." *Proceedings of the National Academy of Sciences USA*, 78:6319–23.

Fisher, R. A. (1930). *The Genetical Theory of Natural Selection*. Oxford: Clarendon.

Flatz, G. (1987). "Genetics of Lactose Digestion in Humans." *Advances in Human Genetics* edited by H. Harris and K. Hirschhorn, 16:1–77.

Fletcher, R. (1991). *Science, Ideology, and the Media: The Cyril Burt Scandal*. New Brunswick, NJ: Transaction.

Flint, J., Hill, A. V. S., Bowden, D. K., Oppenheimer, S. J., Sill, P. R., Serjeantson, S. W., Bana-Koiri, J., Bhatia, K., Alpers, M. P., Boyce, A. J., Weatherall, D. J., and Clegg, J. B. (1986). "High Frequencies of Alpha-Thalassemia Are the Result of Natural Selection by Malaria." *Nature* 321:744–50.

Foucault, M. ([1976] 1978). *The History of Sexuality*, Volume 1. New York: Vintage.

Fraikor, A. L. (1977). "Tay-Sachs Disease: Genetic Drift among the Ashkenazi Jews." *Human Biology* 24:117–34.

Frisancho, R. (1975). "Functional Adaptation to High Altitude Hypoxia." *Science* 187:313–19.

Gajdusek, D. C. (1977). "Unconventional Viruses and the Origin and Disappearance of Kuru." *Science* 197:943–60.

Gajdusek, D. C. (1990). "Raymond Pearl Memorial Lecture, 1989: Cultural Practices as Determinants of Clinical Pathology and Epidemiology of Venereal Infections: Implications for Predictions about the AIDS Epidemic." *American Journal of Human Biology* 2:347–51.

Galton, F. ([1869] 1979). *Hereditary Genius*. London: Julian Friedmann.

Garver, K. L., and Garver, B. (1991). "Eugenics: Past, Present, and Future." *American Journal of Human Genetics* 49:1109–18.

Garver, K. L., and Garver, B. (1994). "The Human Genome Project and Eugenic Concerns." *American Journal of Human Genetics* 54:148–58.

Gates, R. R. (1946). *Human Genetics*. New York: Macmillan.

Geertz, C. (1988). *Works and Lives: The Anthropologist as Author*. Stanford, CA: Stanford University Press.

Gilbert, W. (1992). Pp. 83–97 in *The Code of Codes*, edited by D. J. Kevles and L. Hood. Cambridge, MA: Harvard University Press.

Giles, E., and Elliot, O. (1962). "Race Identification from Cranial Measurements." *Journal of Forensic Sciences* 7:147–57.

Gill, G. W. (1986). "Craniofacial Criteria in Forensic Race Identification." Pp. 143–59 in *Forensic Osteology: Advances in the Identification of Human Remains*, edited by K. Reichs, Springfield, IL: Charles C. Thomas.

Gillispie, C. C. (1951). *Genesis and Geology*. Cambridge, MA: Harvard University Press.

Gilson, E. ([1971] 1984). *From Aristotle to Darwin*. Notre Dame, IN: University of Notre Dame Press.

Ginger, R. (1968). *Six Days or Forever?* Boston: Beacon.

Glass, B., Temkin, O., and Straus, W. L., eds. (1959). *Forerunners of Darwin: 1745–1859*. Baltimore: Johns Hopkins University Press.

Gobineau, A. de ([1854] 1915). *The Inequality of Human Races*, translated by Oscar Levy. London: Heinemann.

Goddard, H. H. (1912). *The Kallikak Family: A Study in the Heredity of Feeble-Mindedness*. New York: Macmillan.

Godfrey, L. R., and Marks, J. (1991). "The Nature and Origins of Primate Species." *Yearbook of Physical Anthropology* 34:39–68.

Goldschmidt, R. B. (1942). "Anthropological Determination of 'Aryanism.'" *Journal of Heredity* 33:215–16.

Goodall, J. (1986). *The Chimpanzees of Gombe: Patterns of Behavior*. Cambridge, MA: Harvard University Press.

Goodman, M. (1962). Immunochemistry of the Primates and Primate Evolution. *Annals of the New York Academy of Sciences*, 102:219–34.

Goodman, M., Tagle, D., Fitch, D. H. A., Bailey, W., Czelusniak, J., Koop, B., Benson, P., and Slightom, J. (1990). "Primate Evolution at the DNA Level and a Classification of Hominoids." *Journal of Molecular Evolution* 30:260–66.

Gould, S. J. (1978). "Morton's Ranking of Races by Cranial Capacity." *Science* 200:503–9.

Gould, S. J. (1981). *The Mismeasure of Man*. New York: Norton.

Gould, S. J. (1982). "The Hottentot Venus." *Natural History* 91(10):20–27.

Gould, S. J. (1983). "Chimp on the Chain." *Natural History* 98(12):18–27.

Gould, S. J., and Lewontin, R. C. (1979). "The Spandrels of San Marco and the Panglossian Paradigm: A Critique of the Adaptationist Programme." *Proceedings of the Royal Society of London* Series B, 205:581–98.

Gould, S. J., and Vrba, E. S. (1982). "Exaptation—A Missing Term in the Science of Form." *Paleobiology* 8:4–15.

Grant, M. (1916). *The Passing of the Great Race*. New York: Scribner's.

Grebner, E. E., and Tomczak, J. (1991). "Distribution of Three a-Chain B-Hexosaminidase A Mutations among Tay-Sachs Carriers." *American Journal of Human Genetics* 48:604–7.

Greene, J. C. (1959). *The Death of Adam*. Ames: Iowa State University Press.

Gregory, W. K. (1913). [Untitled]. *Annals of the New York Academy of Sciences*, 23:268.

Gregory, W. K. (1951). *Evolution Emerging*. New York: Macmillan.

Gregory, W. K., and Hellman, M. (1938). "Evidence of the Australopithecine Man-Apes on the Origin of Man." *Science* 88:615–16.

Grine, F., ed. (1988). *Evolutionary History of the "Robust" Australopithecines*. Hawthorne, NY: Aldine de Gruyter.

Groves, C. P. (1990). "Genes, genitals and genius: The evolutionary ecology of race." Pp. 419–32 in *Human Biology: An Integrative Science, Proceedings of the Fourth Conference of the Australasian Society for Human Biology*, edited by P. O'Higgins.

Hahn, R. A. (1990). "The Meaning of 'Race' and 'Ethnicity' in Government Health Statistics." Paper presented at the American Anthropological Association meetings, December 1990.

Haldane, J. B. S. (1932). *The Causes of Evolution*. London: Longmans, Green.

Haldane, J. B. S. (1940). "The Marxist Philosophy." Pp. 253–75 in *Keeping Cool*. London: Chatto and Windus.

Hall, A. R. (1954). *The Scientific Revolution 1500–1800: The Formation of the Modern Scientific Attitude.* Boston: Beacon.

Hallpike, C. R. (1988). *The Principles of Social Evolution.* Oxford: Clarendon.

Hamer, D., Hu, S., Magnuson, V. L., Hu, N., and Pattatucci, A. M. L. (1993). "A Linkage between DNA Markers on the X Chromosome and Male Sexual Orientation." *Science* 261:321–27.

Handwerker, W. P. (1983). "The First Demographic Transition: An Analysis of Subsistence Choices and Reproductive Consequences." *American Anthropologist* 85:5–27.

Hankins, Frank H. (1926). *The Racial Basis of Civilization: A Critique of the Nordic Doctrine.* New York: Knopf.

Harding, T. S. (1932). "Eugenics for Cows But Not for Humans." *Scientific American* 146 (January):25–27.

Hardy, G. H. (1920). *Some Famous Problems of the Theory of Numbers.* Oxford: Clarendon.

Hardy, G. H. (1940). *A Mathematician's Apology.* Cambridge: Cambridge University Press.

Harris, H. (1966). "Enzyme Polymorphisms in Man." *Proceedings of the Royal Society of London series B,* 164:298–310.

Harris, M. (1968). *The Rise of Anthropological Theory.* New York: Thomas Y. Crowell.

Harrison, G. A. (1993). "Physiologial Adaptation." Pp. 55–72 in *Human Adaptation,* edited by G. A. Harrison. New York: Oxford University Press.

Harrison, G. G. (1975). "Primary Adult Lactase Deficiency: A Problem in Anthropological Genetics." *American Anthropologist* 77:812–35.

Hartwig-Scherer, S., and Martin, R. D. (1991). "Was 'Lucy' More Human Than Her 'Child'? Observations on Early Hominid Postcranial Skeletons." *Journal of Human Evolution* 21:439–49.

Hatch, E. (1973). *Theories of Man and Culture.* New York: Columbia University Press.

Hauser, P., Zametkin, A. J., Martinez, P., Vitiello, B., Matochik, J. A., Mixson, A. J., and Weintraub, B. D. (1993). "Attention Deficit-Hyperactivity Disorder in People with Generalized Resistance to Thyroid Hormone." *New England Journal of Medicine* 328:997–1001.

Hausfater, G., and Hrdy, S. B., eds. (1986). *Infanticide: Comparative and Evolutionary Perspectives.* Hawthorne, NY: Aldine de Gruyter.

Hays, H. R. (1958). *From Ape to Angel.* New York: Capricorn.

Hearnshaw, L. S. (1979). *Cyril Burt, Psychologist.* Ithaca, NY: Cornell University Press.

Hedges, S. B., Kumar, S., Tamura, K., and Stoneking, M. (1992). "Human Origins and Analysis of Mitochondrial DNA Sequences." *Science* 255:737–39.

Henry, W. A. (1993). "Born Gay?" *Time* (July 26):36–39.

Hirschfeld, L., and Hirschfeld, H. (1919). "Serological Differences between the Blood of Different Races." *Lancet* (October 18):675–79.

Herskovits, M. J. (1927). "Variability and Racial Mixture." *American Naturalist* 61:68–81.

Herskovits, M. J. (1938). *Acculturation.* Gloucester, MA: Peter Smith.

Herskovits, M. J. ([1955] 1971) "Cultural Relativism." Pp. 11–34 in Cultural Relativism. New York: Vintage.

Hertzberg, M., Mickelson, K. N. P., Serjeantson, S. W., Prior, J. F., and Trent, R. J. (1989). "An Asian-Specific 9-bp Deletion of Mitochondrial DNA Is Frequently Found in Polynesians." *American Journal of Human Genetics* 44:504–10.

Higgs, D. R., and Weatherall, D. J. (1983). "Alpha-Thalassemia." *Current Topics in Hematology* 4:37–97.

Hill, A., and Ward, S. (1988). "Origin of the Hominidae: The Record of African Large Hominoid Evolution between 14 My and 4 My." *Yearbook of Physical Anthropology* 31:49–83.

Hill, A., Ward, S., Deino, A., Curtis, G., and Drake, R. (1992). "Earliest Homo." *Nature* 355:719–22.

Hilts, P. J. (1992). "Agency Rejects Study Linking Genes to Crime." *New York Times*, September 5.

Hines, M. (1993). "Sex Ratios at Work." *Scientific American* 268(2):12.

Hodgkinson, S., Sherrington, R., Gurling, H., Marchbanks, R., Reeders, S., Mallet, J., McInnes, M., Petursson, H., and Brynjolfsson, J. (1987). "Molecular Genetic Evidence for Heterogeneity in Manic Depression." *Nature* 325:805–6.

Hoffman, E. P. (1994). "The Evolving Genome Project: Current and Future Impact." *American Journal of Human Genetics* 54:129–136.

Hofstadter, R. (1944). *Social Darwinism in American Thought*. Philadelphia: University of Pennsylvania Press.

Hogben, L. (1932). *Genetic Principles in Medicine and Social Science*. New York: Knopf.

Holden, C. (1991). "Alcoholism Gene: Coming or Going?" *Science* 254:200.

Holden, C. (1992). "Twin Study Links Genes to Homosexuality." *Science* 255:33.

Holden, C. (1993). "Hyperactivity Tied to Gene Defect." *Science* 260:295.

Holmes, S. A. (1994). "Federal Government Is Rethinking Its System of Racial Classification." *New York Times*, July 8.

Honig, G. R., and Adams, J. G. III (1986). *Human Hemoglobin Genetics*. New York: Springer-Verlag.

Hook, E. B. (1973). "Behavioral Implications of the Human XYY Genotype." *Science* 179:139–50.

Hooton, E. A. (1926). "Methods of Racial Analysis." *Science* 63:75–81.

Hooton, E. A. (1930a). *The Indians of Pecos Pueblo: Phillips Academy Expedition*, Volume 4. New Haven, CT: Yale University Press.

Hooton, E. A. (1930b). "Doubts and Suspicions Concerning Certain Functional Theories of Primate Evolution." *Human Biology* 2:223–49.

Hooton, E. A. (1931). *Up from the Ape*. New York: Macmillan.

Hooton, E. A. (1936). *Plain Statements about Race*. Science, 83:511-513.

Hooton, E. A. (1939a). *Twilight of Man*. New York: G. P. Putnam's Sons.

Hooton, E. A. (1939b). *The American Criminal: An Anthropological Study*. Volume I: *The Native White Criminal of Native Parentage*. Cambridge, MA: Harvard University Press.

Hooton, E. A. (1946). *Up from the Ape*, 2d edition. New York: Macmillan.

Horai, S., and Hayasaka, K. (1990). "Intraspecific Nucleotide Sequence Differences in the Major Noncoding Region of Human Mitochondrial DNA." *American Journal of Human Genetics* 46:828–42.

Horgan, J. (1992). "D2 or Not D2." *Scientific American* 266(4):29–30.

Horgan, J. (1993). "Eugenics Revisited." *Scientific American* 268(6):122–31.

Horn, J. M., Loehlin, J. L., and Willerman, L. (1979). "Intellectual Resemblance among Adoptive and Biological Relatives: The Texas Adoption Project." *Behavior Genetics* 9:177–207.

Hotchkiss, R. (1979). "The Identification of Nucleic Acids as Genetic Determinants." *Annals of the New York Academy of Sciences* 325:320–42.

Howell, F. C. (1957). "The Evolutionary Significance of Variation and Varieties of 'Neanderthal' Man." *Quarterly Review of Biology* 32:330–410.

Howells, W. W. (1950). "Origin of the Human Stock." *Cold Spring Harbor Symposia on Quantitative Biology* 15:79–85.

Howells, W. W. (1973). *Cranial Variation in Man.* Papers of the Peabody Museum of Archaeology and Ethnology, Harvard University, No. 67.

Howells, W. W. (1992). "Yesterday, Today, and Tomorrow." *Annual Review of Anthropology* 21:1–17.

Hrdy, S. B. (1977). *The Langurs of Abu.* Cambridge: Harvard University Press.

Hrdy, S. B. (1986). "Empathy, polyandry, and the myth of the coy female." Pp. 119–146 in *Feminist Approaches to Science* edited by R. Bleier. New York: Pergamon.

Hrdy, S. B., and Whitten, P. L. (1986). "Patterning of Sexual Activity." Pp. 370–84 in *Primate Societies,* edited by B. B. Smuts, D. L. Cheney, R. M. Seyfarth, R. W. Wrangham, and T. T. Struhsaker. Chicago: University of Chicago Press.

Hubbard, R. (1990). *The Politics of Women's Biology.* New Brunswick, NJ: Rutgers University Press.

Hubbard, R., and Wald, E. (1993). *Exploding the Gene Myth.* Boston: Beacon.

Hughes, D. R. (1968). "Skeletal Plasticity and Its Relevance in the Study of Earlier Populations." Pp. 31–56 in *The Skeletal Biology of Earlier Human Populations,* edited by D. R. Brothwell. Oxford: Pergamon.

Hull, D. L. (1984). "Lamarck among the Anglos." Pp. xi–xvi in *Zoological Philosophy,* J. B. de Lamarck. Chicago: University of Chicago Press

Hulse, F. S. (1955). "Technological Advance and Major Racial Stocks." *Human Biology* 27:184–92.

Hulse, F. S. (1962). "Race as an Evolutionary Episode." *American Anthropologist* 64:929–45.

Hulse, F. S. (1963). *The Human Species.* New York: Random House.

Hunkapiller, T., Huang, H., Hood, L., and Campbell, J. H. (1982). "The Impact of Modern Genetics on Evolutionary Theory." Pp. 1264–89 in *Perspectives on Evolution* edited by R. Milkman. Sunderland, MA: Sinauer.

Hunt, E. E. (1959). "Anthropometry, Genetics, and Racial History." *American Anthropologist* 61:64–87.

Huntington, E. (1924). *The Character of Races.* New York: Scribner's.

Huxley, J. S. (1932). *Problems of Relative Growth.* New York: Dial.

Huxley, J. S. (1938). "Clines: An Auxiliary Taxonomic Principle." *Nature* 142:219–20.

Huxley, J. S., ed. (1940). *The New Systematics.* Oxford: Clarendon.

Huxley, J. S. (1960). "'At Random': A Television Preview." Pp. 41–65 in *Evolution After Darwin*, Volume 3: *Issues in Evolution*, edited by S. Tax. Chicago: University of Chicago Press.

Ingram, V. (1957). "Gene Mutations in Human Hemoglobin: The Chemical Difference between Normal and Sickle Cell Haemoglobin." *Nature* 180:326–28.

Inhorn, M. C., and Brown, P. J. (1990). "The Anthropology of Infectious Disease." *Annual Review of Anthropology* 19:89–117.

Jacob, F. (1977). "Evolution and Tinkering." *Science* 196:1161–66.

Jacobs, P. A., Brunton, M., Melville, M. M., Brittain, R. P., and McClemont, W. F. (1965). "Aggressive Behaviour, Mental Sub-Normality, and the XYY Male." *Nature* 208:1351–52.

Jennings, H. S. (1925). *Prometheus, or Biology and the Advancement of Man*. New York: E. P. Dutton.

Jensen, A. (1992). "Scientific Fraud or False Accusations? The Case of Cyril Burt." Pp. 97–124 in *Research Fraud in the Behavioral and Biomedical Sciences*, edited by D. J. Miller and M. Hersen. New York: Wiley Inter-Science.

Johanson, D. C., Masao, F. T., Eck, G. G., White, T. D., Walter, R. C., Kimbel, W. H., Asfaw, B., Manega, P., Ndessokia, P., and Suwa, G. (1987). "New Partial Skeleton of Homo Habilis from Olduvai Gorge, Tanzania." *Nature* 327:205–9.

Johanson, D. C., and White, T. D. (1979). "A Systematic Assessment of Early African Hominids." *Science* 203:321–29.

Johnson, A. W., and Earle, T. (1987). *The Evolution of Human Societies: From Foraging Group to Agrarian State*. Stanford, CA: Stanford University Press.

Jones, D. (1992). "Sizing-Up the Brain." *Nature* 359:768.

Jones, G. (1980). *Social Darwinism and English Thought*. Atlantic Highlands, NJ: Humanities Press.

Jorde, L. B., and Lathrop, G. M. (1988). "A Test of the Heterozygote-Advantage Hypothesis in Cystic Fibrosis Carriers." *American Journal of Human Genetics* 42:808–15.

Joynson, R. B. (1989). *The Burt Affair*. London: Routledge.

Judson, H. F. (1979). *The Eighth Day of Creation: The Makers of the Revolution in Biology*. New York: Simon and Schuster.

Kamin, L. J. (1986). "Review of Crime and Human Nature, by Wilson and Herrnstein." *Scientific American* 254(2):22–27.

Kan, Y. W., and Dozy, A. (1980). "Evolution of the Hemoglobin S and C Genes in World Populations." *Science* 209:388–91.

Kandel, E. R., and Hawkins, R. D. (1992). "The Biological Basis of Learning and Individuality." *Scientific American* 267(3):78–86.

Kao, F.-T. (1985). "Human Genome Structure." *International Review of Cytology* 96:51–88.

Kaplan, B. (1954). "Environment and Human Plasticity." *American Anthropologist* 56:781–99.

Kelsoe, J. R., Ginns, E. I., Egeland, J. A., Gerhard, D. S., Goldstein, A. M., Bale, S. J., Pauls, D. L., Long, R. T., Kidd, K. K., Conte, G., Housman, D. E., and Paul, S. M. (1989). "Re-Evaluation of the Linkage Relationship between Chromosome 11p Loci and the Gene for Bipolar Affective Disorder in the Old Order Amish." *Nature* 342:238–42.

Kevles, D. J. (1985). *In the Name of Eugenics*. Berkeley: University of California Press.

Kidd, J. R. (1993). "Population Genetics and Population History of Amerindians as Reflected by Nuclear DNA Variation." Ph.D. thesis, Dept. of Anthropology, Yale University.

Kidd, J. R., Kidd, K. K., and Weiss, K. M. (1993). "Human Genome Diversity Initiative." *Human Biology* 65:1–6.

Kimura, D. (1992). "Sex Differences in the Brain." *Scientific American* 267(3):118–25.

Kinzler, K. W. and Vogelstein, B. (1993). "A Gene for Neurofibromatosis 2." *Nature* 363:495–96.

Kitcher, P. (1985). *Vaulting Ambition*. Cambridge: MIT Press.

Klein, A. M. (1991). *Sugarball: The American Game, The Dominican Dream*. New Haven: Yale University Press.

Kleinschmidt, T., and Sgouros, J. G. (1987). "Hemoglobin sequences." *Biological Chemistry Hoppe-Seyler* 368:579–615.

Klineberg, O. (1935). *Race Differences*. New York: Harper and Brothers.

Kluckhohn, C., and Griffith, C. (1950). "Population Genetics and Social Anthropology." *Cold Spring Harbor Symposia in Quantitative Biology* 15:401–8.

Knight, D. (1981). *Ordering the World: A History of Classifying Man*. London: Burnett.

Knowler, W. C., Pettitt, D. J., Bennett, P. H., and Williams, R. C. (1983). "Diabetes Mellitus in the Pima Indians: Genetic and Evolutionary Considerations." *American Journal of Physical Anthropology* 62:107–14.

Konner, M., and Worthman, C. (1980). "Nursing Frequency, Gonadal Function, and Birth Spacing among !Kung Hunter-Gatherers." *Science* 207:788–91.

Koshland, D. E. (1990). "The Rational Approach to the Irrational." *Science* 250:189.

Koshland, D. E. (1993). "Frontiers in Neuroscience." *Science* 262:635.

Kottack, C. P. (1985). "Swimming in Cross-Cultural Currents." *Natural History* 94(5):2–11.

Krause, R. M. (1992). "The Origin of Plagues: Old and New." *Science* 257:1073–78.

Kretchmer, N. (1972). "Lactose and Lactase." *Scientific American* 227(4):70–78.

Krimbas, C. B. (1984). "On Adaptation, Neo-Darwinian Tautology, and Population Fitness." Pp. 1–57 in *Evolutionary Biology*, Volume 17, edited by M. K. Hecht, B. Wallace, and G. T. Prance. New York: Plenum.

Kroeber, A. L. (1916). "Inheritance by Magic." *American Anthropologist* 18:19–40.

Kroeber, A. L. (1917). "The Superorganic." *American Anthropologist* 19:163–213.

Kroeber, A. L. (1923). *Anthropology*. New York: Harcourt, Brace.

Krogman, W. M. (1962). *The Human Skeleton in Forensic Medicine*. Springfield, IL: Charles C. Thomas.

Kühl, S. (1994). *The Nazi Connection: Eugenics, American Racism, and German National Socialism*. New York: Oxford University Press.

Kuhn, T. (1962). *The Structure of Scientific Revolutions*. Chicago: University of Chicago Press.

Kunitz, S. J. (1993). "Diseases and Mortality in the Americas since 1700." Pp.

020-04 in The Cambridge World History of Human Disease, edited by K. F. Kiple. New York: Cambridge University Press.

Kuper, A. (1988). *The Invention of Primitive Society*. New York: Routledge.

Lacoste-Utamsing, C. de, and Holloway, R. L. (1982). "Sexual Dimorphism in the Human Corpus Callosum." *Science* 216:1431–32.

Lamarck, J.-B. ([1809] 1984). *Zoological Philosophy*. Chicago: University of Chicago Press.

Landau, M. (1991). *Narratives of Human Evolution*. New Haven, Yale University Press.

Lang, J. S. (1987). "The Gene Factor: Happiness Is a Reunited Set of Twins." *U.S. News and World Report* (April 13):63–66.

Lapchick, R. (1989). "Pseudo-Scientific Prattle about Athletes." *New York Times*, April 29.

Lapoumeroulie, C., Dunda, O., Ducrocq, R., Trabuchet, G., Mony-Lobe, M., Bodo, J. M., Carnevale, P., Labie, D., Elion, J., and Krishnamoorthy, R. (1992). "A Novel Sickle Cell Mutation of Yet Another Origin in Africa: The Cameroon Type." *Human Genetics* 89:333–37.

Lasch, C. (1991). *The True and Only Heaven: Progress and Its Critics*. New York: Norton.

Lasker, G. (1969). "Human Biological Adaptability." *Science* 166:1480–86.

Layrisse, M. (1958). "Anthropological Considerations of the Diego (Dia)." *Antigen. American Journal of Physical Anthropology* 16:173–86.

Layrisse, M., and Wilbert, J. (1961). "Absence of the Diego Antigen, a Genetic Characteristic of Early Immigrants to South America." *Science* 134:1077–78.

Leaf, A. (1973). "Getting Old." *Scientific American*.

Leakey, L. S. B. (1959). "A New Fossil Skull from Olduvai." *Nature* 184:491–93.

Leakey, L. S. B., Tobias, P. V., and Napier, J. R. (1964). "A New Species of the Genus *Homo* from Olduvai Gorge." *Nature* 202:7–9.

Leakey, M. D., and Hay, R. L. (1979). "Pliocene Footprints in the Laetolil Beds at Laetoli, northern Tanzania." *Nature* 278:317–23.

Lee, R. B., and DeVore, I., eds. (1968). *Man the Hunter*. Chicago: University of Chicago Press.

Lerner, I. M. (1954). *Genetic Homeostasis*. London: Oliver and Boyd.

LeVay, S. (1991). "A Difference in Hypothalamic Structure between Heterosexual and Homosexual Men." *Science* 253:1034–37.

Levi-Strauss, C. ([1962] 1966). *The Savage Mind*. Chicago: University of Chicago Press.

Levins, R., and Lewontin, R. C. (1985). *The Dialectical Biologist*. Cambridge, MA: Harvard University Press.

Lewin, R. (1986). "Shifting Sentiments over Sequencing the Human Genome." *Science* 233:620–21.

Lewontin, R. C. (1972). "The Apportionment of Human Diversity." Pp. 381–98 in *Evolutionary Biology*, Volume 6, edited by M. K. Hecht and W. S. Steere. New York: Plenum.

Lewontin, R. C. (1978). "Adaptation." *Scientific American* 239(3):212–30.

Lewontin, R. C. (1991). "Population Genetics." *Encyclopedia of Human Biology* 6:107–15.

Lewontin, R. C. (1992). "The Dream of the Human Genome." *New York Review of Books* (May 28):31–40.

Lewontin, R. C., and Hubby, J. L. (1966). "A Molecular Approach to the Study of Genic Heterozygosity in Natural Populations. II. Amount of Variation and Degree of Heterozygosity in Natural Populations of Drosophila Pseudoobscura." *Genetics,* 54:595–609.

Lewontin, R. C., Rose, S., and Kamin, L. J. (1984). *Not in Our Genes.* New York: Pantheon.

Linnaeus, C. (1758). *Systema Naturae,* 10th edition. Stockholm: Laurentii Salvii.

Linton, R. (1936). *The Study of Man.* New York: Appleton-Century.

Lippman, W. (1922a). "The Mental Age of Americans." *New Republic* 25 October, pp. 213–15.

Lippman, W. (1922b). "The Mystery of the A Men." *New Republic* 1 November, pp. 246–48.

Lippman, W. (1922c). "The Reliability of Intelligence Tests." *New Republic* 8 November, pp. 275–77.

Lippman, W. (1922d). "Tests of Hereditary Intelligence." *New Republic* 22 November, pp. 328–30.

Lippman, W. (1922e). "A Future for the Tests." *New Republic* 29 November, pp. 9–11.

Little, M. A., and Baker, P. T. (1988). "Migration and Adaptation." Pp. 167–215 in *Biological Aspects of Human Migration,* edited by C. G. N. Mascie-Taylor and G. W. Lasker. New York: Cambridge University Press.

Livingstone, F. B. (1958). "Anthropological Implications of Sickle Cell Gene Distribution in West Africa." *American Anthropologist* 60:533–62.

Livingstone, F. B. (1962). "On the Non-Existence of Human Races." *Current Anthropology* 3:279–81.

Livingstone, F. B. (1963). "Comment on 'Geographic and Microgeographic Races' by Marshall Newman." *Current Anthropology* 4:199–200.

Livshits, G., Sokal, R. R., and Kobyliansky, E. (1991). "Genetic Affinities of Jewish Populations." *American Journal of Human Genetics* 49:131–46.

Loomis, W. F. (1967). "Skin-Pigment Regulation of Vitamin-D Biosynthesis in Man." *Science* 157:501–6.

Lovejoy, A. O. (1936). *The Great Chain of Being.* Cambridge, MA: Harvard University Press.

Ludmerer, K. (1972). *Genetics and American Society: A Historical Appraisal.* Baltimore: Johns Hopkins University Press.

Lukes, S. (1973). *Emile Durkheim: His Life and Work: A Historical and Critical Study.* New York: Penguin.

Luzzatto, L., and Battistuzzi, G. (1985). "Glucose-6-Phosphate Dehydrogenase." Pp. 217–329 in *Advances in Human Genetics,* Volume 14, edited by H. Harris and K. Hirschhorn. New York: Plenum.

Lynn, R. (1982). "IQ in Japan and the United States Shows a Growing Disparity." *Nature* 297:222–23.

Lynn, R. (1992). "Brain Size Differences." *Nature* 359:182.

Lyon, M. F. (1992). "Some Milestones in the History of X-chromosome Inactivation." *Annual Review of Genetics* 26:17–28.

Maddison, D. R. (1991). "African Origin of Human Mitochondrial DNA Revisited." *Systematic Zoology* 40:355–63.

Maddison, D. R., Ruvolo, M., and Swofford, D. L. (1992). "Geographical Origins of Human Mitochondrial DNA: Phylogenetic Evidence from Control Region Sequences." *Systematic Biology* 41:111–24.

Maddox, J. (1992). "How to Publish the Unpalatable?" *Nature*, 358:187.

Maddox, J. (1993). "New Genetics Means No New Ethics." *Nature* 364:97.

Manoach, S. (1992). "The Politics of Finding Homosexuality Genetic." *New York Times*, January 7.

Manoiloff, E. O. (1927). "Discernment of Human Races by Blood: Particularly of Russians from Jews." *American Journal of Physical Anthropology* 10:11–21.

Manoilov, E. O. (1929). "Chemical Reaction of Blood for Definition of Sex in Man, Animals, and Dioecious Plants." *American Journal of Physical Anthropology* 13:29–68.

Marfatia, L., Punales-Morejon, D., and Rapp, R. (1990). "Counseling the Underserved: When an Old Reproductive Technology Becomes a New Reproductive Technology." *Birth Defects* 26:104–26.

Markow, T., Hedrick, P., Zuerlein, K., Danilovs, J., Martin, J., Vyvial, T., and Armstrong, C. (1993). "HLA Polymorphism in the Havasupai: Evidence for Balancing Selection." *American Journal of Human Genetics* 53:943–52.

Marks, J. (1989). "Human Micro- and Macro-Evolution in the Primate Alpha-Globin Gene Family." *American Journal of Human Biology* 1:555–66.

Marks, J. (1991). "What's Old and New in Molecular Phylogenetics." *American Journal of Physical Anthropology* 85:207–19.

Marks, J. (1992a). "Genetic Relationships of the Apes and Humans." *Current Opinion in Genetics and Development* 2:883–89.

Marks, J. (1992b). "Beads and String: The Genome in Evolutionary Theory." Pp. 234–55 in *Molecular Applications in Biological Anthropology*, edited by E. J. Devor. New York: Cambridge University Press.

Marks, J. (1993). "Historiography of Eugenics." *American Journal of Human Genetics* 52:650–52.

Martin. R. (1914). *Lehrbuch der Anthropologie*. Jena: Gustav Fischer.

Mauldin, W. P. (1980). "Population Trends and Prospects." *Science* 209:148–57.

Mayr, E. (1959). "Darwin and the Evolutionary Theory in Biology." Pp. 1–10 in *Evolution and Anthropology: A Centennial Appraisal*, edited by B. J. Meggers. Washington, DC: The Anthropological Society of Washington.

Mayr, E. (1961). "Cause and Effect in Biology." *Science* 134:1501–6.

Mayr, E. (1988). *Toward a New Philosophy of Biology*. Cambridge, MA: Harvard University Press.

Mayr, E. (1994). "Response to John Beatty." *Biology and Philosophy* 9:357–58.

Mazumdar, P. (1992). *Eugenics, Human Genetics and Human Failings: The Eugenics Society, Its Sources and Its Critics in Britain*. New York: Routledge.

McCarty, M. (1985). *The Transforming Principle: Discovering that Genes Are Made of DNA*. New York: Norton.

McClintock, B. (1950). "The Origin and Behavior of Mutable Loci in Maize." *Proceedings of the National Academy of Sciences* 36:344–55.

McClintock, B. (1956). "Controlling Elements and the Gene." *Cold Spring Harbor Symposia on Quantitative Biology* 21:197–216.

McCracken, R. D. (1971). "Lactase Deficiency: An Example of Dietary Evolution." *Current Anthropology* 12:479–517.

McEvedy, C. (1988). "The Bubonic Plague." *Scientific American* 258(2):118–23.

McEwen, J. E., McCarty, K., and Reilly, P. R. (1993). "A Survey of Medical Directors of Life Insurance Companies Concerning Use of Genetic Information." *American Journal of Human Genetics* 53:33–45.

McEwen, J. E., and Reilly, P. R. (1992). "State Legislative Efforts to Regulate Use and Potential Misuse of Genetic Information." *American Journal of Human Genetics* 51:637–47.

McKeown, T. (1988). *The Origins of Human Disease*. Cambridge, MA: Basil Blackwell.

McKusick, V. A., Eldridge, R., Hostetler, R. A., and Egeland, J. A. (1964). "Dwarfism in the Amish." *Transactions of the Association of American Physicians* 77:151–68.

McNeill, W. (1976). *Plagues and Peoples*. Garden City, NY: Doubleday.

Medawar, P. ([1963] 1991). Is the Scientific Paper a Fraud? Pp. 228–33 in *The Threat and the Glory: Reflections on Science and Scientists*. New York: Oxford University Press.

Medawar, P. ([1965] 1984)."Two Conceptions of Science." Pp. 28–41 in *Pluto's Republic*. New York: Oxford University Press.

Meindl, R. S. (1987). "Hypothesis: A Selective Advantage for Cystic Fibrosis Heterozygotes." *American Journal of Physical Anthropology* 74:39–45.

Mencken, H. L. (1927). "On Eugenics." *Baltimore Sun*, May 15.

Merbs, C. (1992). "A New World of Infection Disease." *Yearbook of Physical Anthropology* 35:1–42.

Merriwether, D. A., Clark, A. G., Ballinger, S. W., Schurr, T. G., Soodyall, H., Jenkins, T., Sherry, S. T., and Wallace, D. C. (1991). "The Structure of Human Mitochondrial DNA Variation." *Journal of Molecular Evolution* 33:543–55.

Merson, M. (1993). "Slowing the Spread of HIV: Agenda for the 1990s." *Science* 260:1266–68.

Merton, R. K., and Montagu, M. F. A. (1940). "Crime and the Anthropologist." *American Anthropologist* 42:384–408.

Michael, J. S. (1988). "A New Look at Morton's Craniological Research." *Current Anthropology* 29:348–54.

Miringoff, M.-L. (1991). *The Social Costs of Genetic Welfare*. New Brunswick, NJ: Rutgers University Press.

Molleson, T., Jones, K., and Jones, S. (1993). "Dietary Change and Effects of Food Preparation on Microwear Patterns in the Late Neolithic of Abu Hureyra, Northern Syria." *Journal of Human Evolution* 24:455–68.

Montagu, A. (1963). "What Is Remarkable about Varieties of Man Is Likenesses, Not Differences." *Current Anthropology* 4:361–64.

Montagu, A., ed. (1964). *Man's Most Dangerous Myth: The Fallacy of Race*, 4th edition. Cleveland: World.

Montagu, M. F. A. (1941). "The Concept of Race in the Human Species in the Light of Genetics." *Journal of Heredity* 32:243–47.

Montagu, M. F. A. (1943). "Edward Tyson, M.D., F.R.S. 1650–1708 and the Rise of Human and Comparative Anatomy in England." *Memoirs of the American Philosophical Society* 20.

Morell, V. (1993). "Evidence Found for a Possible 'Aggression Gene.'" *Science* 260:1722–23.

Morgan, T. H. (1924). "Human Inheritance." *American Naturalist* 58:385–409.

Morgan, T. H. (1925). *Evolution and Genetics*. Princeton, NJ: Princeton University Press.

Morphy, E. (1994). "Cultural Adaptation." Pp. 99–150 in *Human Adaptation*, edited by G. A. Harrison. New York: Oxford University Press.

Morris, D. (1967). *The Naked Ape*. New York: McGraw-Hill.

Motulsky, A. (1960). "Metabolic Polymorphisms and the Role of Infectious Diseases in Human Evolution." *Human Biology* 32:28–62.

Motulsky, A. (1989). "Societal Problems in Human and Medical Genetics." *Genome* 31:870–75.

Mourant, A. E. (1983). *Blood Relations: Blood Groups and Anthropology*. New York: Oxford University Press.

Mourant, A. E., Kopeç, A. C., and Domaniewska-Sobczak, K. (1976). *The Distribution of The Human Blood Groups*, 2d edition. London: Oxford University Press.

Mourant, A. E., Kopeç, A. C., and Domaniewska-Sobczak, K. (1978). *The Genetics of the Jews*. Oxford: Clarendon.

Muller, H. J. (1933). "The Dominance of Economics over Eugenics." *Scientific Monthly* 37:40–47.

Müller-Hill, B. (1988). *Murderous Science*. New York: Oxford University Press.

Myerowitz, R. (1988). "Splice Junction Mutation in Some Ashkenazi Jews with Tay-Sachs Disease: Evidence against a Single Defect within This Ethnic Group." *Proceedings of the National Academy of Sciences* 85:3955–59.

Myrianthopoulos, N. C., and Aronson, S. M. (1966). "Population Dynamics of Tay-Sachs Disease. I. Reproductive Fitness and Selection." *American Journal of Human Genetics* 18:313–27.

Myrianthopoulos, N. C., and Aronson, S. M. (1972). "Population Dynamics of Tay-Sachs Disease. II. What Confers the Selective Advantage upon the Jewish Heterozygote?" Pp. 561–70 in *Sphingolipids, Sphingolipidoses, and Allied Disorders*, edited by B. W. Volk and S. M. Aronson. New York: Plenum.

Myrianthopoulos, N. C., Naylor, A. F., and Aronson, S. M. (1972). "Founder Effect in Tay-Sachs Disease Unlikely." *American Journal of Human Genetics* 24:341–42.

Napier, J. ([1980] 1993). *Hands*. Princeton, NJ: Princeton University Press.

Natowicz, M. R., Alper, J. K., and Alper, J. S. (1992). "Genetic Discrimination and the Law." *American Journal of Human Genetics* 50:465–75.

Navon, R., and Proia, R. L. (1989). "The Mutations in Ashkenazi Jews with Adult GM2 Gangliosidosis, the Adult Form of Tay-Sachs Disease." *Science* 243:1471–74.

Neel, J. V. (1949). "The Inheritance of Sickle Cell Anemia." *Science* 110:64–66.

Neel, J. V. (1962). "Diabetes Mellitus: A 'Thrifty' Genotype Rendered Detrimental by 'Progress'?" *American Journal of Human Genetics*, 14:353–62.

Nei, M., and Roychoudhury, A. (1981). "Genetic Relationship and Evolution of Human Races." Pp. 1–59 in *Evolutionary Biology*, Volume 14, edited by M. K. Hecht, B. Wallace, and G. T. Prance. New York: Plenum.

Nei, M., and Roychoudhury, A. (1993). "Evolutionary Relationships of Human Populations on a Global Scale." *Molecular Biology and Evolution* 10:927–43.

Newman, H. H. (1932). *Evolution, Genetics, and Eugenics*, 3d edition. Chicago: University of Chicago Press.

Newman, M. T. (1953). "The Application of Ecological Rules to the Racial Anthropology of the Aboriginal New World." *American Anthropologist* 55:311–27.

Newmark, P. (1986). "Is Megasequencing Madness?" Nature, 323:291.

Nisbet, R. (1980). *History of the Idea of Progress*. New York: Basic Books.

Nitecki, M., ed. (1988). *Evolutionary Progress? Chicago: University of Chicago Press.*

O'Brien, M., and Holland, T. D. (1990). "Variation, Selection, and the Archaeological Record." Pp. 31–79 in *Archaeological Method and Theory*, Volume 2, edited by M. B. Schiffer. University of Arizona Press, Tucson.

O'Brien, S., ed. (1993). *Genetic Maps: Locus Maps of Complex Genomes*, 6th edition. Cold Spring Harbor: Cold Spring Harbor Laboratory Press.

O'Brien, S. J., Wildt, D. E., and Bush, M. (1986). "The Cheetah in Genetic Peril." *Scientific American* 254(5):84–92.

O'Brien, S. J., Wildt, D. E., Goldman, D., Merril, C. R., and Bush, M. (1983). "The Cheetah Is Depauperate in Genetic Variation." *Science* 221:459–62.

Ohno, S. (1970). *Evolution by Gene Duplication*. New York: Springer-Verlag.

Olshansky, S. J., Carnes, B. A., and Cassel, C. K. (1993). "The Aging of the Human Species." *Scientific American* 268(4):46–52.

Olsen, J. (1968). "The Black Athlete, Part 1: The Cruel Deception." *Sports Illustrated* (July 1):15–27.

Olson, M. V. (1993). "The Human Genome Project." *Proceedings of the National Academy of Sciences* 90:4338–44.

Ottenberg, R. (1925). "A Classification of Human Races Based on Geographic Distribution of the Blood Groups." *Journal of the American Medical Association* 84:1393–95.

Padover, S., ed. (1978). *The Essential Marx*. New York: Mentor.

Parker, L. S. (1994). "Bioethics for Human Geneticists: Models for Reasoning and Methods for Teaching." *American Journal of Human Genetics* 54:137–47.

Paley, W. (1802). *Natural Theology*. London: R. Fauldner.

Pauling, L. (1968). "Reflections on the New Biology." *UCLA Law Review* 15(3):269.

Pauling, L., Itano, H., Singer, S. J., and Wells, I. C. (1949). "Sickle Cell Anemia, a Molecular Disease." *Science* 110:543–48.

Pearl, R. (1927). "The Biology of Superiority." *American Mercury* 12:257–66.

Pearson, K. (1909). *On the Scope and Importance to the State of the Science of National Eugenics*. London: Dulau.

Pearson, K. (1900). "Race Crossing in Jamaica." *Nature* 126.427–29.

Perna, N. T., Batzer, M. A., Deininger, P. L., and Stoneking, M. (1992). "Alu Insertion Polymorphism: A New Type of Marker for Human Population Studies." *Human Biology* 64:641–48.

Piot, P., Plummer, F. A., Mhalu, F. S., Lamboray, J.-L., Chin, J., and Mann, J. M. (1988). "AIDS: An International Perspective." *Science* 239:574–79.

Poliakov, L. (1974). *The Aryan Myth: A History of Racist and Nationalist Ideas in Europe*. New York: Meridian.

Poliakowa, A. T. (1927). "Manoiloff's 'Race' Reaction and Its Application to the Determination of Paternity." *American Journal of Physical Anthropology* 10:23–29.

Ponder, B. (1990). "Neurofibromatosis Gene Cloned." *Nature* 346:703–4.

Pool, R. (1993). "Evidence for Homosexuality Gene." *Science* 261:291–92.

Popper, K. (1963). *Conjectures and Refutations*. New York: Basic Books.

Proctor, R. N. (1988). *Racial Hygiene: Medicine under the Nazis*. Cambridge, MA: Harvard University Press.

Proctor, R. N. (1991). *Value-Free Science? Purity and Power in Modern Knowledge*. Cambridge, MA: Harvard University Press.

Provine, W. B. (1971). *The Origins of Theoretical Population Genetics*. Chicago: University of Chicago Press.

Provine, W. B. (1973). "Geneticists and the Biology of Race Crossing." *Science* 182:790–96.

Provine, W. B. (1986). *Sewall Wright and Evolutionary Biology*. Chicago: University of Chicago Press.

Putnam, C. (1961). *Race and Reason*. Washington, DC: Public Affairs Press.

Quetel, C. ([1986] 1990). *History of Syphilis*. Baltimore: Johns Hopkins University Press.

Quevedo, W. C., Jr., Fitzpatrick, T. B., and Jimbow, K. (1985). "Human Skin Color: Origin, Variation and Significance." *Journal of Human Evolution* 14:43–56.

Rapp, R. (1988). "Chromosomes and Communication: The Discourse of Genetic Counseling." *Medical Anthropology Quarterly* 2:143–57.

Rapp, R. (1990). "Constructing Amniocentesis: Maternal and Medical Discourses." Pp. 28–42 in *Uncertain Terms: Negotiating Gender in American Culture*, edited by Faye Ginsburg and Anna Lowenhaupt Tsing. Boston: Beacon.

Reed, T. E. (1969). "Caucasian Genes in American Negroes." *Science* 762–68.

Reichenbach, H. (1951). *The Rise of Scientific Philosophy*. Berkeley,CA: University of California Press.

Rennie, J. (1994). "Grading the Gene Tests." *Scientific American* 270(6):88–97.

Resta, R. G. (1992). "The Twisted Helix: An Essay on Genetic Counselors, Eugenics, and Social Responsibility." *Journal of Genetic Counseling* 1:227–43.

Reus, V. I. (1993). "Mind and Brain." *Science* 262:1629.

Rice, Thurman B. (1929). *Racial Hygiene*. New York: Macmillan.

Rindos, D. (1985). "Darwinian Selection, Symbolic Variation, and the Evolution of Culture." *Current Anthropology* 26:65–88.

Ripley, W. Z. (1899). *The Races of Europe*. New York: Appleton.

Risch, N. (1992). "Genetic Linkage: Interpreting Lod Scores." *Science* 255:803–4.

Risch, N., Squires, B., Wheeler, E., and Keats, B. J. (1993). "Male Sexual Orientation and Genetic Evidence." *Science* 262:2063–64.

Robarchek, C. (1990). "Motivations and Material Causes: On the Explanation of Conflict and War." Pp. 56–76 in *The Anthropology of War*, edited by J. Haas. New York: Cambridge University Press.

Roberts, C. W. M., Shutter, J. R., and Korsmeyer, S. J. (1994). "Hox11 Controls the Genesis of the Spleen." *Nature* 368:747–749.

Roberts, D. F. (1953). "Body Weight, Race, and Climate." *American Journal of Physical Anthropology* 11:533–58.

Roberts, D. F. (1963). "Review of The Origin of Races." *Human Biology* 35:443–45.

Roberts, D. F. (1981). "Culture and Disease in Britain and Western Europe." Pp. 417–58 in *Biocultural Aspects of Disease*, edited by H. Rothschild. Orlando, FL: Academic Press.

Roberts, L. (1991a). "A Genetic Survey of Vanishing Peoples." *Science* 252:1614–17.

Roberts, L. (1991b). "Genetic Survey Gains Momentum." *Science* 254:517.

Roberts, L. (1992a). "The Huntington's Gene Quest Goes On." *Science* 258:740–41.

Roberts, L. (1992b). "Genome Diversity Project: Anthropologists Climb (Gingerly) on Board." *Science* 258:1300–1.

Robertson, M. (1987). "Molecular Genetics of the Mind." *Nature* 325:755.

Robertson, M. (1989). "False Start on Manic Depression." *Nature* 342:222.

Robins, R. S. (1981). "Disease, Political Events and Populations." Pp. 153–75 in *Biocultural Aspects of Disease*, edited by H. Rothschild. Orlando, FL: Academic Press.

Rogers, J. A. (1993). "A Population Genetics Perspective on the Human-Chimpanzee-Gorilla Trichotomy." *Journal of Human Evolution* 25:201–15.

Rogers, S. L. (1984). *The Human Skull: Its Mechanics, Measurements, and Variations*. Springfield, IL: Charles C. Thomas.

Rosenberg, A. (1985). *The Structure of Biological Science*. Cambridge, MA: MIT Press.

Rothenberg, R. B. (1993). "Gonorrhea." Pp. 756–63 in *The Cambridge World History of Human Disease*, edited by K. F. Kiple. New York: Cambridge University Press.

Rousseau, J.-J. ([1755] 1984). *A Discourse on Inequality*. New York: Penguin.

Rowe, C. (1950). "Genetics vs. Physical Anthropology in Determining Racial Types." *Southwestern Journal of Anthropology* 6:197–211.

Ruano, G., Rogers, J. A., Fergusoin-Smith, A. C., and Kidd, K. K. (1992). "DNA Sequence Polymorphism within Hominoid Species Exceeds the Number of Phylogenetically Informative Characters for a HOX2 Locus." *Molecular Biology and Evolution* 9:575–86.

Rudwick, M. J. S. (1976). *The Meaning of Fossils*, 2d edition. Chicago: University of Chicago Press.

Ruffie, J. (1986). *The Population Alternative* . New York: Pantheon.

Ruse, M. (1988). *Taking Darwin Seriously*. New York: Basil Blackwell.

Rushton, J. P. (1992a). "Differences in Brain Size." *Nature* 358:532.

Rushton, J. P. (1992b). "The Brain Size/IQ Debate." *Nature* 360:292.

Russell, J. (1993). "No Way to Treat a Body." *New Scientist* (3 July):24–28.

Rusting, R. L. (1992). "Why Do We Age?" *Scientific American* 267(6):130–41.

Ryan, F. (1993). *The Forgotten Plague: How the Battle against Tuberculosis Was Won—and Lost.* Boston: Little, Brown.

Saavedra, J. M., and Perman, J. A. (1991). "Lactose Malabsorption and Intolerance." *Encyclopedia of Human Biology* 4:611–21.

Sachar, H. M. (1992). *A History of the Jews in America.* New York: Vintage.

Scarr, S., and Weinberg, R. A. (1978). "Attitudes, Interests, and IQ." *Human Nature* 1:29–36.

Schiebinger, L. (1993). *Nature's Body.* Boston: Beacon.

Schluter, D. (1992). "Brain Size Differences." *Nature* 359:182.

Schmid, C. W., and Jelinek. W. R. (1982). "The Alu Family of Dispersed Repetitive Sequences." *Science* 216:1065–70.

Schmid, C. W., and Shen, C.-K. J. (1985). "The Evolution of Interspersed Repetitive DNA Sequences in Mammals and Other Vertebrates." Pp. 323–58 in *Molecular Evolutionary Genetics,* edited by R. J. MacIntyre. New York: Plenum.

Schull, W. J. (1993). "Ethnicity and Disease—More Than Familiality." *American Journal of Human Biology* 5:373–85.

Schurr, T. G., Ballinger, S. W., Gan, Y.-Y., Horge, J. A., Merriwether, D. A., Lawrence, D. N., Knowler, W. C., Weiss, K. M., and Wallace, D. C. (1990). "American Mitochondrial DNAs Have Rare Asian Mutations at High Frequencies, Suggesting They Derived from Four Primary Maternal Lineages." *American Journal of Human Genetics* 46:613–23.

Scozzari, R., Torroni, A., Semino, O., Sirugo, G., Brega, A., and Santachiara-Benerecetti, A. S. (1988). "Genetic Studies on the Senegal Population." I. Mitochondrial DNA Polymorphisms. *American Journal of Human Genetics* 43:534–44.

Scriven, M. (1959). "Explanation and Prediction in Evolutionary Theory." *Science* 130:477–82.

Scriver, C. R., Kaufman, S., and Woo, S. L. C. (1988). "Mendelian Hyperphenylalanenemia." *Annual Review of Genetics* 22:301–21.

Scully, D. G., Dawson, P. A., Emmerson, B. T., and Gordon, R. B. (1992). "A Review of the Molecular Basis of Hypoxanthine-Guanine Phosphoribosyltransferase (HPRT) Deficiency." *Human Genetics* 90:195–207.

Seligmann, J., and Foote, D. (1991). "Whose Baby Is It, Anyway?" *Newsweek* (October 28):73.

Selvin, P. (1991). "The Raging Bull of Berkeley." *Science* 251:368–71.

Serjeant, G. R. (1992). *Sickle Cell Disease, 2d edition.* New York: Oxford University Press.

Shapiro, H. L. (1939). *Migration and Environment.* New York: Oxford University Press.

Shapiro, H. L. (1961). "Race Mixture." Pp. 333–89 in *The Race Question in Modern Science.* New York: Columbia University Press.

Shatz, C. J. (1992). "The Developing Brain." *Scientific American* 267(3):60–67.

Shea, C. (1994). "Queens College and a Measure of Diversity." *Human Diversity* 90:195–207.

Shen, M. R., Batzer, M. A., and Deininger, P. L. (1991). "Evolution of the Master Alu Gene(s)." *Journal of Molecular Evolution* 33:311–20.

Shipman, P. (1994). *The Evolution of Racism*. New York: Simon and Schuster.

Shipman, P., Walker, A., and Bichell, D. (1985). *The Human Skeleton*. Cambridge, MA: Harvard University Press.

Simon, M. A. (1978). "Sociobiology: The Aesop's Fables of Science." *Sciences* 16:18–21, 31.

Simpson, G. G. (1963). "Biology and the Nature of Science." *Science* 139:81–88.

Singer, C. (1957). *A Short History of Anatomy and Physiology from the Greeks to Harvey*. New York: Dover.

Slack, P. (1989). "The Response to Plague in Early Modern England: Public Policies and Their Consequences." Pp. 167–87 in *Famine, Disease and the Social Order in Early Modern Society*, edited by J. Walter and R. Schofield. New York: Cambridge University Press.

Snyder, L. H. (1926). "Human Blood Groups: Their Inheritance and Racial Significance." *American Journal of Physical Anthropology* 9:233–63.

Snyder, L. H. (1930). "The 'Laws' of Serologic Race-Classification Studies in Human Inheritance IV." *Human Biology* 2:128–33.

Sokal, R. R. (1991a). "Ancient Movement Patterns Determine Modern Genetic Variances in Europe." *Human Biology* 63:589–606.

Sokal, R. R. (1991b). "The Continental Population Structure of Europe." *Annual Review of Anthropology* 20:119–40.

Solomon, E. (1990). "Colorectal Cancer Genes." *Nature* 343:412–14.

Soloway, R. A. (1990). *Demography and Degeneration: Eugenics and the Declining Birthrate in Twentieth-Century Britain*. Chapel Hill, NC: University of North Carolina Press.

Solway, J., and Lee, R. B. (1990). "Foragers, Genuine or Spurious?" *Current Anthropology* 31:109–46.

Spencer, H. (1852). "The Development Hypothesis." *The Leader* (March 20):280–81.

Spencer, H. (1896). *Principles of Sociology*, 3d edition. New York: Appleton.

Spier, L. (1956). "Inventions and Human Society." Pp. 224–46 in *Man, Culture, and Society*, edited by H. L. Shapiro. New York: Oxford University Press.

Spyropoulos, B. (1988). "Tay-Sachs Carriers and Tuberculosis Resistance." *Nature* 331:666.

Spyropoulos, B., Moens, P. B., Davidson, J., and Lowden, J. A. (1981). "Heterozygote Advantage in Tay-Sachs Carriers?" *American Journal of Human Genetics* 33:375–80.

Stanbridge, E. J. (1990). "Identifying Tumor Suppressor Genes in Human Colorectal Cancer." *Science* 247:12–13.

Stanton, W. (1960). *The Leopard's Spots*. Chicago: University of Chicago Press.

Stern, C. (1943). "The Hardy-Weinberg Law." *Science* 97:137–38.

Stewart, T. D. (1951a). "Scientific Responsibility." *American Journal of Physical Anthropology* 9:1–3.

Stewart, T. D. (1951b). "Objectivity in Racial Classifications." *American Journal of Physical Anthropology* 9:470–72.

Stinson, S. (1992). "Nutritional Adaptation." *Annual Review of Anthropology* 21:143–70.

Stocking, G. W. (1968). *Race, Culture, and Evolution: Essays in the History of Anthropology.* New York: Free Press.

Stocking, G. W. (1987). *Victorian Anthropology.* New York: Free Press.

Stoneking, M, Sherry, S. T., and Vigilant, L. (1992). "Geographic Origin of Human Mitochondrial DNA Revisited." *Systematic Biology* 41:384–91.

Stout, J. T., and Caskey, C. T. (1985). "Lesch-Nyhan Syndrome." *Annual Review of Genetics* 19:127–48.

Stout, J. T., and Caskey, C. T. (1988). "The Lesch-Nyhan Syndrome: Clinical, Molecular and Genetic Aspects." *Trends in Genetics* 4:175–78.

Strandskov, H. H., and Washburn S. L. (1951). "Genetics and Physical Anthropology." *American Journal of Physical Anthropology* 9:261–63.

Streuver, S, ed. (1971). *Prehistoric Agriculture.* Garden City, NY: Natural History Press.

Stringer, C. B., and Andrews, P. (1988). "Genetic and Fossil Evidence for the Origin of Modern Humans." *Science* 239:1263–68.

Sumner, W. G. ([1906] 1940). *Folkways.* Boston: Ginn.

Susman, R. L. (1988). "Hand of Paranthropus Robustus from Member 1, Swartkrans: Fossil Evidence for Tool Behavior." *Science* 240:781–84.

Susman, R. L., Stern, J. T., and Jungers, W. L. (1984). "Arboreality and Bipedality in the Hadar Hominids." *Folia primatologica* 43:113–56.

Suzuki, D., and Knudtson, P. (1990). *Genethics: The Clash between the New Genetics and Human Values.* Cambridge MA: Harvard University Press.

Swaab, D. F., and Fliers, E. (1985). "A Sexually Dimorphic Nucleus in the Human Brain." *Science* 228:1112–15.

Taylor, G. (1921). "The Evolution and Distribution of Race, Culture, and Language." *Geographical Review* 11:54–119.

Teasdale, T. W., and Owen, D. R. (1984). "Heredity and Familial Environment in Intelligence and Educational Level—A Sibling Study." *Nature* 309:620–22.

Teilhard de Chardin, P. (1959). *The Phenomenon of Man.* New York: Harper and Row.

Temin, H., and Engels, W. (1984). "Movable Genetic Elements and Evolution." Pp. 173–201 in *Evolutionary Theory: Paths into the Future,* edited by J. W. Pollard. New York: Wiley.

Templeton, A. R. (1992). "Human Origins and Analysis of Mitochondrial DNA Sequences." *Science* 255:737.

Templeton, A. R. (1993). "The 'Eve' Hypothesis: A Genetic Critique and Reanalysis." *American Anthropologist* 95:51–72.

Terman, L. M. (1916). *The Measurement of Intelligence.* Boston: Houghton Mifflin.

Thieme, F. P. (1952). "The Population as a Unit of Study." *American Anthropologist* 54:504–9.

Thomson, G. (1983). "The Human Histocompatibility System: Anthropological Considerations." *American Journal of Physical Anthropology* 62:81–89.

Thorne, A. G., and Wolpoff, M. H. (1981). "Regional Continuity in Australasian Pleistocene Hominid Evolution." *American Journal of Physical Anthropology* 55:337–59.

Thornhill, R., and Thornhill, N. (1983). "Human Rape: An Evolutionary Analysis." *Ethology and Sociobiology* 4:137–73.

Thornhill, R., and Thornhill, N. (1992). "The Evolutionary Psychology of Men's Coercive Sexuality." *Behavioral and Brain Sciences* 15:363–421.

Tobias, P. V. (1970). "Brain Size, Grey Matter and Race—Fact or Fiction?" *American Journal of Physical Anthropology* 32:3–26.

Tobias, P. V. (1978). "The Life and Work of Linnaeus." *South African Journal of Science* 74:457–62.

Tobias, P. V. (1992). *Olduvai Gorge,* Volume 4: *The Skulls, Endocasts, and Teeth of Homo habilis).* New York: Cambridge University Press.

Trabuchet, G., Chebloune, Y., Savatier, P., Lachuer, J., Faure, C., Verdier, G., and Nigon, V. M. (1987). "Recent Insertion of an Alu Sequence in the Beta-Globin Gene Cluster of the Gorilla." *Journal of Molecular Evolution* 25:288–91.

Trautmann, T. R. (1992). "The Revolution in Ethnological Time." *Man* 27:379–97.

Travathan, W. (1987). *Human Birth: An Evolutionary Perspective.* Hawthorne, NY: Aldine de Gruyter.

Travis, J. (1993). "New Tumor Suppressor Gene Captured." *Science* 260:1235.

Trent, J. W., Jr. (1994). *Inventing the Feeble Mind.* Berkeley: University of California Press.

Trivers, R. (1972). *Parental Investment and Sexual Selection.* In: Sexual Selection and the Descent of Man, 1871–1971, edited by Bernard Campbell. Chicago: Aldine.

Tsai, F. C. S. (1992). "Brain Size Differences." *Nature* 359:182.

Tsui, L.-C., and Buchwald, M. (1991). "Biochemical and Molecular Genetics of Cystic Fibrosis." Pp. 153–266 in *Advances in Human Genetics,* Volume 20, edited by H. Harris and K. Hirschhorn. New York: Plenum.

Tyler, E. B. (1871). *Primitive Culture.* London: John Murray.

Tyson, E. ([1699] 1966). *Orang-Outang, sive Homo sylvestris: or, the Anatomy of a Pygmie.* London: Dawsons of Pall Mall.

Vacher de Lapouge, G. (1896). *Les Sélections Sociales.* Paris: A. Fontemoing.

Vacher de Lapouge, G. (1899). *L'Aryen: Son Role Social.* Paris: A. Fontemoing.

Vallois, H. V. (1953). "Race." Pp. 145–62 in *Anthropology Today,* edited by A. L. Kroeber. Chicago: University of Chicago Press.

Van Valen, L. M. (1974). "Brain Size and Intelligence in Man." *American Journal of Physical Anthropology* 40:417–23.

Van Valen, L. M. (1992). "Sizing-Up the Brain." *Nature* 359:768.

Vigilant, L., Stoneking, M., Harpending, H., Hawkes, K., and Wilson, A. C. (1991). "African Populations and the Evolution of Human Mitochondrial DNA." *Science* 253:1503–07.

Waddington, C. H. (1942). "Canalization of Development and the Inheritance of Acquired Characters." *Nature* 150:563–65.

Waddington, C. H. (1957). *The Strategy of the Genes.* London: Allen and Unwin.

Waddington, C. H. (1960). "Evolutionary Adaptation." Pp. 381–402 in *Evolution after Darwin,* Volume I: *The Evolution of Life,* edited by S. Tax. Chicago: University of Chicago Press.

Wallace, B. (1958). "The Average Effect of Radiation-Induced Mutations on Viability in Drosophila Melanogaster." *Evolution* 12:532–56.

Wallace, D. (1992). *The Search for the Gene*. Ithaca, NY: Cornell University Press.

Wallace, M. R., Andersen, L. B., Saulino, A. M., Gregory, P. E., Glover, T. W., and Collins, F. S. (1991). "A de novo Alu Insertion Results in Neurofibromatosis Type 1." *Nature* 353:864–66.

Walsh, J. B., and Marks, J. (1986). "Sequencing the Human Genome." *Nature* 322:590.

Walter, H. E. ([1913] 1932). *Genetics*. New York: Macmillan (reprinted in Newman, 1932).

Walters, L. (1988). "Ethical Issues in the Prevention and Treatment of HIV Infection and AIDS." *Science* 239:597–603.

Ward, R. (1983). "Genetic and Sociocultural Components of High Blood Pressure." *American Journal of Physical Anthropology* 62:91–105.

Washburn, S. L. (1951). "The New Physical Anthropology." *Transactions of the New York Academy of Sciences* Series II, 13:298–304.

Washburn, S. L. (1963). "The Study of Race." *American Anthropologist* 65:521–31.

Watson, J. D. (1990). "The Human Genome Project: Past, Present, and Future." *Science* 248:44–51.

Weidenreich, F. (1945). "The Brachycephalization of Recent Mankind." *Southwestern Journal of Anthropology* 1:1–54.

Weidenreich, F. (1947). "Facts and Speculations Concerning the Origin of Homo Sapiens." *American Anthropologist* 49:135–51.

Weinberger, A. D. (1964). "A Reappraisal of the Constitutionality of 'Miscegenation' Statutes." Pp. 402–24 in *Man's Most Dangerous Myth: The Fallacy of Race*, 4th edition, edited by A. Montagu. Cleveland: World.

Weiner, A. M., Deininger, P. L., and Efstradiatis, A. (1986). "Nonviral Retroposons: Genes, Pseudogenes and Transposable Elements Generated by the Reverse Flow of Genetic Information." *Annual Review of Biochemistry* 55:631–61.

Weiss, K. M., Ferrell, R. E., and Hanis, C. L. (1984). "A New World Syndrome of Metabolic Diseases with a Genetic and Evolutionary Basis." *Yearbook of Physical Anthropology* 27:153–78.

Weiss, K. W., Kidd, K. K., and Kidd, J. R. (1992). "Human Genome Diversity Project." *Evolutionary Anthropology* 1:80–82.

Weiss, R. A. (1993). "How Does HIV Cause AIDS?" *Science,* 260:1273–79.

Westoff, C. F. (1986). "Fertility in the United States." *Science* 234:554–59.

White, L. A. (1947). "The Expansion of the Scope of Science." *Journal of the Washington Academy of Sciences* 37:181–210 (reprinted in White, 1949).

White, L. A. (1949). *The Science of Culture*. New York: Farrar, Straus, and Giroux.

Wiener, A. S. (1945). "Rh-Hr Blood Types in Anthropology." *Yearbook of Physical Anthropology* 1:212–14.

Wiener, A. S. (1948). "Blood Grouping Tests in Anthropology." *American Journal of Physical Anthropology* 6:236–37.

Wiggins, D. K. (1989). "'Great Speed but Little Stamina': The Historical Debate over Black Athletic Superiority." *Journal of Sports History* 16:158–85.

Wiley, E. O. (1981). *Phylogenetics: The Theory and Practice of Phylogenetic Systematics*. New York: Wiley.

Willard, H. F. (1991). "Evolution of Alpha Satellite." *Current Opinion in Genetics and Development* 1:509–14.

Williams, G. C. (1966). *Adaptation and Natural Selection.* Princeton, NJ: Princeton University Press.

Wilmsen, E. (1989). *Land Filled with Flies: A Political Economy of the Kalahari.* Chicago: University of Chicago Press.

Wilson, E. O. (1978). *On Human Nature.* Cambridge, MA: Harvard University Press.

Wilson, J. Q., and Herrnstein, R. J. (1985). *Crime and Human Nature.* New York: Simon and Schuster.

Witkin, H. A., Mednick, S. A., Schulsinger, F., Bakkestrom, E., Christiansen, K. O., Goodenough, D. R., Hirschhorn, K., Lundsteen, C., Owen, D. R., Philip, J., Rubin, D. R., and Stocking, M. (1976). "Criminality in XYY and XXY Men." *Science* 193:547–55.

Wolf, E. R. (1982). *Europe and the People without History.* Berkeley: University of California Press.

Wolpoff, M. H., Wu Xinzhi, and Thorne, A. G. (1984). "Modern Homo Sapiens Origins: A General Theory of Hominid Evolution Involving the Fossil Evidence from East Asia." Pp. 411–83 in *The Origins of Modern Humans: A World Survey of the Fossil Evidence*, edited by F. H. Smith and F. Spencer. New York: Alan R. Liss.

Wood, B. A. (1991). *Koobi Fora Research Project, Volume 4: Hominid Cranial Remains.* New York: Oxford University Press.

Wood, B. A. (1992). "Origin and Evolution of the Genus Homo." *Nature* 355:783–90.

Wood, E. J., and Bladon, P. T. (1985). *The Human Skin.* London: Edward Arnold.

Woods, F. A. (1923). "A Review of Reviews of Madison Grant's Passing of the Great Race." *Journal of Heredity* 14:93–95.

Wright, L. (1994). "One Drop of Blood." *New Yorker* (July 25):46–55.

Wright, S. (1931). "Evolution in Mendelian Populations." *Genetics* 16:97–159.

Wright, S. (1932). "The Roles of Mutation, Inbreeding, Cross-Breeding, and Selection in Evolution." *Proceedings of the Sixth International Congress of Genetics* 1:356–66.

Wright, S. (1986). *Evolution: Selected Papers.* Chicago: University of Chicago Press.

Wrischnik, L. A., Higuchi, R. G., Stoneking, M., Erlich, H. A., Arnheim, N., and Wilson, A. C. (1987). "Length Mutations in Human Mitochondrial DNA; Direct Sequencing of Enzymatically Amplified DNA." *Nucleic Acids Research* 15:529–42.

Yerkes, R. M., ed. (1921). "Psychological Examining in the U. S. Army." *Memoirs of the National Academy of Sciences* 15. Washington, DC: National Academy of Sciences.

Young, J. (1989). "Making Good." *Nature* 342:138.

Zinsser, H. (1935). *Rats, Lice and History.* Boston: Little, Brown.

Index